# The end of telecoms history

*How we reached the point of having all that we need*

*William Webb*

The end of telecoms history

ISBN-13: 9798328402729

© William Webb, 2024

# Table of Contents

Structure of this book ................................................................................... 1
1    Why the end of history? ........................................................................ 3
    1.1    Francis Fukuyama ........................................................................ 3
    1.2    Fibre take-up is well below 100% ............................................... 4
    1.3    5G has disappointed .................................................................... 6
    1.4    Smartphones fail to innovate ....................................................... 9
    1.5    An industry in pain ...................................................................... 9
    1.6    Summary ................................................................................... 11
2    A very abridged history of telecoms ................................................... 12
    2.1    Introduction ............................................................................... 12
    2.2    Fixed ......................................................................................... 12
    2.3    Mobile ....................................................................................... 16
    2.4    Broadcast .................................................................................. 22
    2.5    Shannon .................................................................................... 25
    2.6    Geopolitics ................................................................................ 26
    2.7    Becoming a utility .................................................................... 27
    2.8    Wrapping up history ................................................................. 30
3    User requirements are nearly met ........................................................ 31
    3.1    Introduction ............................................................................... 31
    3.2    The Internet .............................................................................. 31
    3.3    The iPhone effect ..................................................................... 35
    3.4    Video drives usage ................................................................... 37
    3.5    Data confirms predictions ........................................................ 39
        3.5.1    Mobile data growth figures .............................................. 39
        3.5.2    Fixed data growth figures ................................................ 45
    3.6    Latency is sufficient, faster is prohibitively expensive ............ 48
    3.7    We have all we need ................................................................ 49
4    Is there anything that could reignite growth? ..................................... 50
    4.1    Moving demand requires something huge ............................... 50
    4.2    The 5G hopefuls ....................................................................... 50
    4.3    AR/VR ...................................................................................... 55
    4.4    AI .............................................................................................. 57
    4.5    IoT and digital twins ................................................................ 58

|      |     |                                              |     |
|------|-----|----------------------------------------------|-----|
|      | 4.6 | FWA                                          | 61  |
|      | 4.7 | Autonomous cars                              | 63  |
|      | 4.8 | Paying for growth                            | 64  |
|      | 4.9 | A solution in search of a problem            | 65  |
| 5    |     | Implications                                 | 67  |
|      | 5.1 | Introduction                                 | 67  |
|      | 5.2 | 6G and Wi-Fi8                                | 67  |
|      | 5.3 | Manufacturers                                | 69  |
|      | 5.4 | Operators                                    | 70  |
|      | 5.5 | Academics                                    | 74  |
|      | 5.6 | Regulators                                   | 75  |
|      | 5.7 | Politicians                                  | 77  |
|      | 5.8 | The industry as a whole                      | 85  |
| 6    |     | Delivering ubiquity                          | 86  |
|      | 6.1 | Coverage is not improving                    | 86  |
|      | 6.2 | There are new tools – HAPs and D2D           | 88  |
|      | 6.3 | Subsidy will be needed                       | 91  |
|      | 6.4 | Shared networks                              | 93  |
|      | 6.5 | Trains                                       | 94  |
|      | 6.6 | In-building                                  | 95  |
|      | 6.7 | A change of focus from speed to coverage     | 97  |
|      | 6.8 | Fixed networks                               | 98  |
|      | 6.9 | Delivering ubiquity                          | 99  |
| 7    |     | Conclusions                                  | 100 |
|      | 7.1 | In summary                                   | 100 |
|      | 7.2 | Satiation is a good thing                    | 101 |
|      | 7.3 | New thinking is needed                       | 102 |

# List of abbreviations

| | |
|---|---|
| ADSL | Asynchronous Digital Subscriber Line |
| AI | Artificial Intelligence |
| AR | Augmented Reality |
| CDMA | Code Division Multiple Access |
| D2D | Direct to Device (satellite to handset) |
| DTMF | Dual Tone Multi-Frequency (signalling) |
| DTTV | Digital Terrestrial Television |
| DVD | Digital Video Disk |
| eMBB | Enhanced Mobile Broadband |
| FTTC | Fibre to the Cabinet |
| FTTH | Fibre to the Home |
| FWA | Fixed Wireless Access |
| GPRS | General Packet Radio Service |
| HAPs | High Altitude Platforms |
| HD | High Definition (video) |
| HSPA | High Speed Packet Access |
| IoT | Internet of Things |
| IPR | Intellectual Property Right |
| ISDN | Integrated Services Digital Line |
| ITU | International Telecommunications Union |
| MIMO | Multiple Input Multiple Output (antenna) |
| MMC | Massive Machine Communications |
| MMS | Mobile Message Service |
| MNO | Mobile Network Operator |
| MVNO | Mobile Virtual Network Operator |
| OFDM | Orthogonal Frequency Division Multiplexing |
| O-RAN | Open Radio Access Network |
| OTT | Over the top |
| PON | Passive Optical Network |
| QAM | Quadrature Amplitude Modulation |
| RAN | Radio Access Network |
| SA | Standalone (5G network) |
| SMS | Short Message Service |
| SNR | Signal to Noise Ratio |

| | |
|---|---|
| SRN | Shared Rural Network |
| URLLC | Ultra-reliable low latency communications |
| VCR | Video Cassette Recorder |
| VoLTE | Voice over LTE |
| VR | Virtual Reality |
| XR | Extended Reality |

# Structure of this book

Many assume that the growth in fixed and mobile data usage will continue indefinitely and indeed is "exponential"[1]. There is also a widespread view that we need ever-faster "gigabit" connectivity, and that new applications that will require better networks are just around the corner. This book shows how these view are completely wrong.

Instead, the growth in data usage is slowing and will soon plateau at a point where we are consuming all the data we need. Beyond data rates of around 10-20Mbits/s there are no discernible benefits as applications are slowed by other factors such as server latency. Current networks in much of the world deliver more than enough data and speed. We have no need of anything better. And there are no applications likely to change this in the foreseeable future – which is at least ten, perhaps twenty years. With sufficiency of connectivity the end of telecoms history has arrived.

The remaining challenge is ubiquity – ensuring that we have this connectivity everywhere – to all homes and wherever people take their mobile phone. This is less a technical challenge and more an economic one.

This book sets out the case for why history has come to an end for telecoms. It then looks at the implications and the remaining task of delivering ubiquity.

1. Chapter 1 sets out the reasons for concluding that we are at the end of history covering recent developments such as the disappointment of 5G.
2. Chapter 2 provides a very abridged and selective history of telecoms to date, showing how it has, until recently, focused on finding solutions to clear problems, mostly insufficient speed and capacity, but in the last decade has moved towards being a solution looking for a problem.
3. Chapter 3 sets out the case as to why user requirements for connectivity have now been met, showing that we have greater speeds than we can

---

[1] Which, to be pedantic, means not only is usage growing but the rate of that growth is increasing.

use and that growth rates for data usage are falling and will reach zero in the next few years.
4. Chapter 4 asks whether there are any applications that might change the situation, looking at whether 5G and 6G proposed applications are likely and would materially change demand, data rates and latency. It concludes that, just as with 5G, the 6G proposed applications are unlikely.
5. Chapter 5 addresses the implications of the end of history on manufacturers, operators, regulators, academics, politicians and others in the supply chain.
6. Chapter 6 turns to the remaining task – of delivering sufficiency of communications everywhere, discussing ways to enhance mobile coverage and broadband connectivity in not-spots and under-served areas.
7. Chapter 7 delivers a summary and conclusions.

# 1 Why the end of history?

## *1.1 Francis Fukuyama*

Francis Fukuyama famously published a book in 1992 entitled "The end of history and the last man"[2]. The main thesis was that as countries around the world moved to liberal democracies there was no reason for more debate on the best way to organise society. The history, at least of how we best organise our societies, was over. This was based on data showing ever-increasing number of countries moving towards democracy and the expectation that this trend would inexorably continue.

Since then, the trend towards liberal democracies has been reversed and much history has happened. Dictators, or those who would clearly like to become dictators, have emerged in many countries.

This is not a book about democracy, but it draws an analogy to the concept that once a trend is settled and inevitable, that there is little need for more "history" – for innovations, change, experimentation and unexpected developments. This book makes the case that this is so in telecommunications: that we now have all of the connectivity we need, apart from in "not spot" areas, and as a result there is no need to develop anything more – there is no more history that need be written.

Francis Fukuyama was proven wrong. The history of our politics and societies has not come to an end, quite the contrary over the last two decades. Predicting telecoms, though, is much easier, and indeed I have a strong track record of doing so accurately over the last 25 years. For example, I published "The Future of Wireless Communications" in 2001 where I set out a clear set of predictions for communications in 2020[3]. These predictions proved almost exactly correct, including estimates of home broadband rates, device form

---

[2] https://www.amazon.co.uk/End-History-Last-Francis-Fukuyama/dp/0743284550
[3] See, W T Webb, "The Future of Wireless Communications", Artech House, 2001, W T Webb, "Wireless Communications: The Future", Wiley, Chichester, 2007, W Webb, "Our Digital Future", Amazon, October 2017.

factors and likely applications. The evidence presented here is very compelling. And borrowing a phrase from Fukuyama communicates well the key message of the book.

The rest of this section sets the scene for why we should consider telecommunications to have reached the end of its century-long growth period, after which the issues are considered in more detail in separate chapters.

## 1.2 Fibre take-up is well below 100%

Many governments have made the provision of universal fibre to every premise a key priority. They have subsidised deployments or in some cases, including Australia and Ireland, set up national entities charged with reaching full fibre.

Fibre is seen as important because of the virtually unlimited bandwidth it provides. With legacy copper solutions, data rates rarely exceed around 100Mbits/s and in rural areas can be below 10Mbits/s. Fibre can easily deliver 1Gbits/s and with appropriate upgrades 10 or even 100Gbits/s is perfectly possible.

Yet, despite subsidised deployments, and this apparent 10x-100x increase in improvements, fibre take-up is low in many countries. The charts below show both coverage – the number of homes that have access to fibre – and penetration rates – the percentage of homes that have access to fibre that choose to change from their existing broadband to take up the newly available fibre solution. in the EU and UK

Penetration levels are growing steadily but take up is below 50% across Europe and only around 30% in the UK. Why, if fibre is such a good thing, is it not adopted as soon as it becomes available?

## Why the end of history?

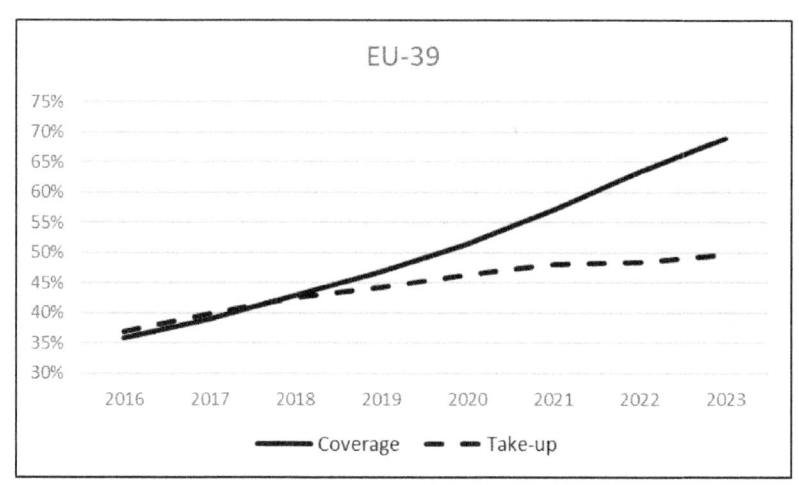

Figure 1-1: FTTH penetration across Europe: Source FTTH Council [4]

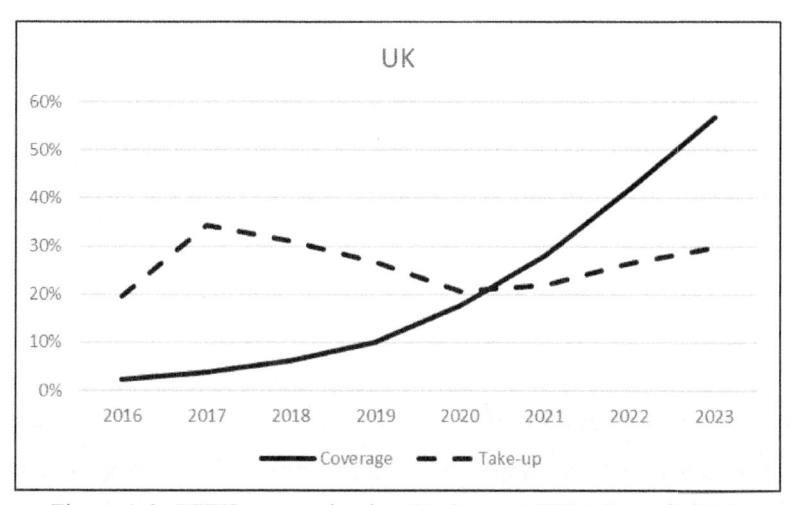

Figure 1-2: FTTH penetration in UK: Source FTTH Council (ibid)

There are some barriers to adoption. While fibre may pass by the house, there can be disruption at bringing it from the street into the home, including the need to trench across the front garden or driveway and install an active termination point within the home. Fibre solutions are also often slightly more expensive,

---

[4] https://www.ftthcouncil.eu/resources/all-publications-and-assets/2154/2024-ftth-market-panorama-report-by-country

albeit typically only 10-20%, than legacy solutions. But neither of these would be problematic if the benefits promised of unlimited home and business connectivity were seen as valuable.

The data does not seem available to know exactly why each homeowner has not decided to take up fibre. But it seems highly plausible that many are content with the data rates that they currently have[5]. As will be explained in more detail in Chapter 3, a rate of 10Mbits/s per person in the building would allow each person to be watching a high-definition video stream simultaneously and still have spare bandwidth. With most homes having two to three inhabitants the rates provided by legacy copper are, for many, more than sufficient. Homeowners see no reason, nor are incentivised, to upgrade to something that is even just a little more expensive. Sufficiency is at the heart of the argument made in this book.

In passing, upgrading to fibre to nearly all premises is a sensible strategy at a national level – but more because it reduces network maintenance costs, lowers power consumption and improves reliability rather than because it delivers end-user benefits. I say "nearly all premises" because for the more rural dwellings the costs of deploying fibre can be prohibitive and other solutions, considered in Chapter 6 are more appropriate. Because of these other benefits, fibre solutions will eventually be the same cost or less than existing copper solutions and households will migrate on the basis of cost and perhaps reliability. Hence, penetration levels will rise over time.

## 1.3  5G has disappointed

5G was supposed to be "a generation like no other". CTOs of major mobile network operators (MNOs) gushed that 5G would be more transformative than the advent of electricity. Adverts showed remote surgery from the back of a church during a wedding.

---

[5] This is despite blatant mis-selling in some cases with adverts suggesting that 1Gbits/s is needed for everyone in a household to be on-line simultaneously.

## Why the end of history?

I warned about this in my book "The 5G Myth"[6]. Published in 2016, about four years before the first substantive 5G deployments, it set out why the features that were being offered would not be of much interest. Hence, it predicted that 5G would not lead to increased revenues for the MNOs and as a result the business case would be poor. MNOs would be worse off as a result of deploying it, and consumers no better off.

I will return later to what users actually need, but in summary, 5G promises three key feature sets, as follows:

1. *Enhanced mobile broadband (eMBB).* This would deliver higher data rates than 4G – taking rates from around 20Mbits/s to well over 200Mbits/s. However, once mobile users have data rates of around 5Mbits/s then the speed at which apps run is limited by other factors such as server latency. With 5Mbits/s high-definition video is possible (and hardly needed on small handset screens) and internet access tends to be constrained by other factors such as far end servers. Broadly, very few will notice the benefits.
2. *Massive machine connectivity (MMC).* The intent here was to enable Internet of Things (IoT) deployments at much higher densities than currently possible. However, IoT solutions are only slowly being deployed and current 4G technologies such as NB-IoT are more than sufficient. And, as it happens, 5G did not implement anything significantly new for IoT in any case.
3. *Ultra-low latency reliable communications (URLLC).* This was a mixed concept - lower latency was intended to facilitate new applications that required real-time remote control, while the reliability was intended to make 5G suitable for industrial applications. However, the market for such solutions is very small, and can often be met by fibre or Wi-Fi deployments. And the network upgrade needed is expensive. It is only in private 5G deployments – in a few specific areas such as factories or mines where the low-latency benefits appear worthwhile, but this is a very small market compared to the global mobile marketplace.

---

[6] W Webb, "The 5G Myth", First edition: Amazon, November 2016. Second edition: Amazon, December 2017. Third edition: DeGruyter Press, January 2019.

Some four years into 5G deployments – which is nearly halfway through its anticipated leadership before 6G arrives – much of what I predicted in "The 5G Myth" has come true. Consumers have not been impressed, for example Deloitte reported[7]:

> "As familiarity with 5G smartphones grows, consumers are developing more realistic expectations regarding the capabilities and performance of 5G networks. In 2022, 48% of 5G smartphone users we surveyed reported that the service exceeded their expectations; in 2023, this percentage has declined to 38%. As 5G networks expand and mature, users' expectations will likely align more closely with the capabilities of the technology."

This suggests that for the majority 5G is not meeting expectations. IoT subscriptions to MNOs remain low and there are few if any commercial instances of URLLC.

The reason why 5G has not been "a generation like no other" is because most of the features it provided were not needed. It was a solution in search of a problem – and no problem existed. Consumers, where they had good 4G connections, had all the mobile connectivity that they needed to do whatever they wanted with their phones. Having more was of no benefits. IoT deployments likewise had all they needed.

Like fibre, there is some sense in 5G. More capacity has been needed for mobile networks as data usage grew significantly – a topic that will be a major part of this book with a section devoted to discussing it in Chapter 3. While that capacity could have been delivered using 4G, 5G technology is somewhat better at providing capacity and hence appropriate for MNOs to deploy in areas where networks need more capacity, which is broadly what most have done.

5G was predicated on the hope of "build it and they will come". "They" have not come and are highly unlikely to do so.

---

[7] https://www2.deloitte.com/us/en/insights/industry/telecommunications/connectivity-mobile-trends-survey/2023/future-of-5g-and-challenges.html

## 1.4 Smartphones fail to innovate

Until around 2010 handsets changed dramatically each year. New form factors emerged, screens improved, some phones had keyboards, others folded. But in 2007 the iPhone launched and within a few years all phones followed the same model – a screen taking up all of the front of the phone and a camera on the back.

Phones are now marketed not for their communications capabilities, but for their cameras, the quality of their screens and factors such as robustness and battery life. And phones have hardly changed over the last decade. A few foldable phones have emerged but failed to gain widespread adoption. Beyond that, it has been marginal gains in camera pixels, device memory and processor speed. The processor speed may be important in areas such as running AI-powered personal advisors (such as Siri) but this is not related to communications.

This is not necessarily a bad thing – it suggests that we have reached a form factor and level of capability that suits the majority. Phones do all that we want them to, and there is little need for a better one in order to improve communications.

## 1.5 An industry in pain

Back in the 1990s, the telecoms industry was one of the fastest growing. MNOs quickly entered the top financial indices and had one of the highest levels of profitability. Equipment manufacturers such as Motorola, Ericsson, Nokia and Nortel saw rapidly growing markets.

Fast forward to the 2020s and the situation could not be more different. MNOs have had a terrible time with shares dramatically under-performing the market over the last decade. Many manufacturers no longer exist having been acquired or split apart. MNOs are claiming that they are in such poor financial health that the over-the-top (OTT) providers such as Google, Amazon, Apple and others should subsidise them. Job cuts in both manufacturers and operators are rampant.

The chart below shows the performance of European telecommunication stocks versus the average European market index[8].

**Europe Telecoms Versus General Index**

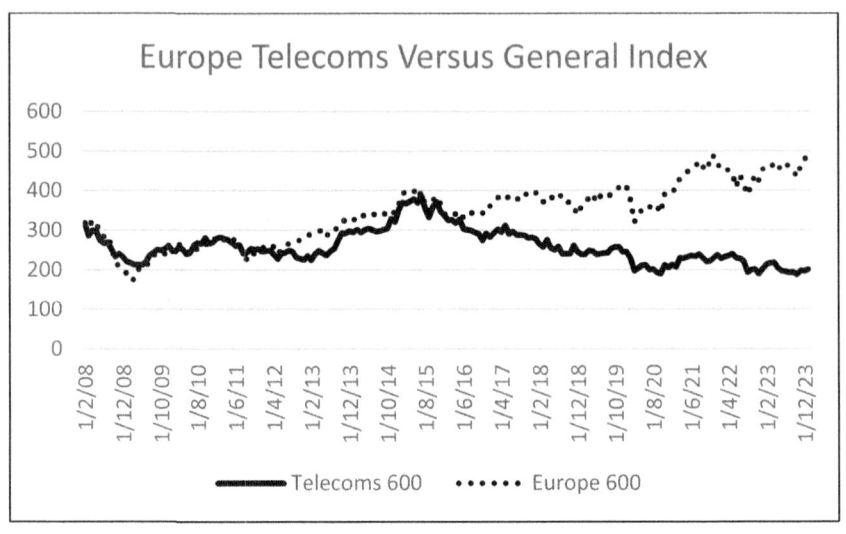

Figure 1-3: MNO share prices relative to the index

Ever since 2000 they have underperformed and this is particularly significant since 2015 where, broadly telecoms shares have declined by about 50% while the market overall grew by around 65%. The effect is that an investment of $1 in telecoms shares in 2015 would now be worth around $0.65 while the same investment in a general tracker fund would be worth $1.65. Telecoms has been a terrible investment.

In a paper entitled "Financing Broadband Networks of the Future"[9] and packed with valuable data, the OECD assessed various measures of financial performance of operators and noted:

> The performance of communication operators displays notable divergence across various indices. European operators experienced a decline of 47% from 1 January 2008 to 30 June 2023. Conversely, the

---

[8] It shows the Stoxx 600 Telecoms versus the Stoxx 600 Europe
[9] https://www.oecd.org/digital/financing-broadband-networks-of-the-future-eafc728b-en.htm

## Why the end of history?

MSCI World Telecom Services Index, encompassing developments in 23 developed markets, increased by 8%. Meanwhile, communication operators in the United States, as measured by the S&P Telecom Select Industry, rose by 33% over the past 15 years. In comparison, the broader stock market, represented by the MSCI World Index, experienced an overall growth of 87%, while the Dow Jones surged by 159% over the same period.

Some of this pain is self-imposed, with the industry setting an inappropriate direction for 5G and then making investments with poor payback. But at heart it stems from almost everyone (with some very significant exceptions I will return to) having all the fixed and mobile connectivity that they need. Hence, there is neither growth in subscriber numbers nor in willingness to pay. The result has been that operators have not seen revenue growth, or at least not real growth when taking inflation into account, but they have seen increased expenditure on 5G spectrum and networks, or fibre deployment. If an industry chooses to make bad decisions, that is its own problem, and it should not be rescued as a result. But it is a symptom of a lack of interest in connectivity beyond that already available with 4G.

I will return to those that have insufficient connectivity in Chapter 6. This includes those in rural areas in most countries and those who cannot afford connectivity in developing countries. Resolving this is the remaining challenge for connectivity and I will set out ways to rise to this challenge in the later part of the book.

## *1.6 Summary*

In this introduction I have pointed to a number of factors that suggest that we have all the communications that we need. I will show why this is so in more detail in later chapters, but first, to explain why this means we have reached the end of telecoms history, I will set out some of the history that has brought us to this point.

## 2  A very abridged history of telecoms

### 2.1  Introduction

There are many excellent resources covering the history of telecommunications, and little need to repeat them here. Instead, what I want to show is that for most of history, the telecoms industry was trying to catch up with demand. Developments were solutions to clear problems. But in recent years, the industry has switched to offering solutions in search of problems.

That history could be considered to start in 1844 when Samuel Morse sent his first message and saw a major advance in 1901 when Guglielmo Marconi made the first trans-Atlantic wireless transmission.

While there continues to be convergence between fixed, mobile and broadcast systems, they remain sufficiently discrete to be considered here in separate subsections.

### 2.2  Fixed

Fixed communications (as in being connected via a wire, cable or fibre) was the first to emerge. The clear preference was solutions that allowed people to speak to each other, but initial telegraph systems did not have the bandwidth to allow this. Instead, a code had to be invented – Morse code – that allowed messages to be laboriously relayed. It took from 1844 to around 1878 before the first workable telephone systems were implemented, until 1915 for the first coast-to-coast call in the US, and until 1927 for the first trans-Atlantic call.

The first 100 years of the phone – from the 1870s to the 1970s – were very much focused on delivering voice calls to everyone. Deploying telephone lines to all homes across the developed world was a major undertaking. Ever better undersea cables were needed to handle the desire for international calls. Exchanges had to become larger and then automated in order to handle call volumes. Broadly, it was an era of solving a clear problem – enabling anyone in their home or business to speak to anyone in any other home or business, anywhere in the world. Quality improved and prices declined as technology became progressively better.

A key advance was the introduction of dual-tone multi-frequency (DTMF) signalling. This sent short dual-tone "beeps" down the phone line and allowed rotary phones to be replaced with push-button phones that were easier to use and enabled larger, faster and more efficient exchanges. And perhaps more importantly, it heralded the way for data to be sent over the telephone system.

By the 1970s the voice problem had broadly been solved – at least in developed countries. The next development was to allow the phone system to deliver documents, replacing post and couriers. This was achieved with the facsimile or "fax" machine. The fax machine scans a document and converts each line into a digital representation – essentially a binary coding of whether there is black or white paper under each pixel in the scanner. The binary data is then fed down the phone line, using the same bandwidth as a voice call, and the fax machine at the far end acts as a printer, printing black ink wherever there was a black pixel in the original.

The speed of transmission was limited by the bandwidth of the voice line, meaning it could take many minutes to send longer documents. The fax machine used a standard voice line, but then that line could not be used for a voice call during the process of sending a fax and the recipient would need to know to set the fax machine to answer the incoming call rather than answer it as a voice call. For that reason, businesses generally had lines dedicated to the fax machine.

The fax machine solved a need for a better way to transfer documents, especially where this needed to be done faster than possible with the postal service. It did require a bulky machine that had to be kept maintained with ink and paper and the resulting print-out was typically not as clear as the original, especially for pictures and complex images.

The next need emerged with the invention of the Internet. While the seeds were laid in the 1970s, it was not until around 1989 that the first commercial connections to the Internet occurred and not until the mid-1990s that the various elements were in place to make it a useful consumer and business prospect. The arrival of the Internet led to a growing demand to be connected to it. Initially this was via fixed networks – both because mobile networks were not sufficiently advanced and also because interactions with the Internet until around 2005 were almost exclusively through a PC or similar.

The initial solutions to connecting to the Internet were similar to those used for the fax machine – sending a digital signal down the phone line within the bandwidth normally used for voice calls. And just like the fax, this prevented a voice call at the same time. The means of connecting a computer to a phone line was a "modem" – a modulator/demodulator – which converted the signal from the computer into a form suitable for the telephone line. Early models were slow – initially around 1200bps (or 1.2kbits/s). As can be appreciated knowing the rates we have now, this was painfully slow. A 1Mbyte attachment would have taken almost two hours to download. There was growing demand for faster transmission and various technologies such as quadrature amplitude modulation (QAM) were employed to increase rates. Rates increased to 9.6kbits/s in 1984, 19.2kbits/s in 1993 and up to the heady heights of 56kbits/s by 1998, although this could only be achieved on lines with low noise levels. For many, typical speeds were around 33kbits/s.

These rates were clearly still far too slow. Logging on and retrieving emails was hugely time-consuming, especially if there were attachments, and any kind of video was nearly impossible. There was huge demand for something faster. Meeting this demand has been the story of the evolution of fixed telecoms ever since.

The restriction on going faster was the bandwidth of the phone line which had been developed to handle voice calls. These did not need to transmit frequencies above around 8kHz. Two approaches emerged:

1. Deploy different types of lines that were optimised for data.
2. Remove the 8kHz restriction on the voice network.

Both were followed and both continue to be used. For some homes there were existing cable connections implemented to distribute broadcast content. Because these had been designed for TV bandwidths which were hugely higher than

## A very abbreviated history of telecoms

voice, one of the "TV channels" could be re-purposed for data, delivering much higher rates while still allowing broadcast content to be received.

An initial solution to removing the 8kHz restriction was to deploy Integrated Services Digital Network (ISDN) lines which implemented digital voice, and so could "take over" the entire line. ISDN delivered 64kbits/s - or double that if two lines could be used together. But ISDN was not that much faster than the modems then in use and required different telephones and was not widely adopted.

The most widely adopted solution emerged in the late 1990s with asymmetric digital subscriber line (ADSL) technology. This aimed to allow voice calls to continue in their existing bandwidth but to use higher frequencies for data. This required changes to the network and special sockets inside the home that acted as "splitters" to separate the voice and data networks. Original ADSL modems introduced from 1998 could achieve 8Mbits/s downlink speeds and 1Mbits/s uplink although the actual rate delivered was hugely reliant on the quality of the copper phone line. Longer lines to homes further from the local exchange could only manage much lower rates.

ADSL was a huge improvement. Data rates were over 100 times faster than dial-up modems and the voice line could be used at the same time. ADSL was progressively improved with ADSL2+ delivering up to 24Mbits/s by 2005. However, with many not being able to achieve these rates, there was still significant demand for faster connectivity to access the video content and larger attachments that were becoming more commonplace on the Internet.

Going faster required the increasing use of fibre into the network. Fibre delivered direct to a home enables data rates of well over 100Mbits/s but replacing the entire copper network which had been deployed over many decades was a huge task, and alternative solutions were sought. Instead, fibre was taken from the local exchange to the street cabinets – known as "fibre to the cabinet" (FTTC). Modems were then placed in the cabinet to use the remaining short stretches of copper to the home. These "VDSL" modems could deliver up to 300Mbits/s by 2016, although few lines were of sufficient quality to enable this. Instead, rates of around 50Mbits/s were typically achieved – but this was a huge improvement on the rates actually delivered with ADSL which were often less than 10Mbits/s.

That brings us to the present day, where most homes in the developed world have, as a minimum, VDSL service in the region of 50Mbits/s, and many have cable modems and fibre connections that can deliver 1Gbits/s or more. Only those in more remote areas have lower rates because of the length of copper line still used, and the cost of deploying dedicated fibre or cable connections.

My thesis, which I will be developing across fixed, mobile and broadcast in this chapter, and then in much greater detail in subsequent chapters, is that until recently fixed networks were evolving in response to a clear problem, or demand. It was obvious initially that people would want to speak to each other without being in the same place. The need for faster document transmission was obvious from the costs of couriers widely used to speed documents around cities. And the need for Internet connectivity became obvious as the Internet rose to prominence. Initial data rates were clearly insufficient, leading to ever-greater efforts to deliver sufficient fast connectivity. Faster connectivity in turn spurred new applications such as video to become available, further increasing connectivity demands. Since the mid-1990s, engineers have been continually trying to keep up. But there is no reason why this race to sufficiency should go on forever. Eventually we will have enough bandwidth, new applications will no longer emerge, and we can turn attention elsewhere.

As I will argue in subsequent chapters, I believe that point has arrived. We can now send multiple 4k video streams via existing home broadband. Users are not, generally, demanding faster connectivity as evidenced in the low penetration rates of fibre solutions. But politicians and others still seem convinced that the "race" continues and that the nice round number of 1Gbits/s is an appropriate end goal. Such a goal is, currently, a solution in search of a problem, unlike all prior fixed telecoms history where the problems have been all too clear, and solutions have emerged to resolve them.

## 2.3  Mobile

Mobile phones started out as car phones. Indeed, the name Motorola – one of the pioneers – came from the company's initial focus on delivering radios into cars hence "Motor", and the "ola" was a suffix used on some musical instruments such as a pianola. Motorola then branched out into delivering two-way radios, initially for police and then for taxis and other similar businesses. Even by the 1980s the car was still the main focus for the mobile phone – one of the UK's biggest retailers of mobile phones still went by the name "Carphone Warehouse" in the 2020s.

Just as with fixed lines, voice calls were the initial focus. In the 1930s there were major restrictions to voice calls. The first was that the two-way radios were bulky and expensive – too big to carry and only suited to being vehicle mounted. The second was that there was not much capacity in the radio systems implemented at the time, meaning that only a small number of people could make a call at any one time.

The challenge of size and weight took until the 1970s and the advent of transistors and better batteries to start to resolve, and it was not until the 1980s that phones started to become portable. The challenge of capacity took a similar length of time to resolve. In December 1947, Bell Labs engineers Douglas Ring and W. Rae Young proposed hexagonal cell transmissions for mobile phones. But the technology to implement such a system did not exist then and the radio frequencies had not yet been allocated. It took until 1973 for Motorola manager Marty Cooper to place what is often called the first cellphone call.

The 1G era of mobile phones from around 1982 to 1992, saw increasingly widespread deployment of analogue mobile phone solutions that were designed only for voice – the only requirement at the time. Quickly it became clear that 1G would not have the capacity for a level of demand much greater than predicted. As 1G was increasingly widely adopted it also became clear there were many other problems:

- Most critical was security. This allowed conversations to be readily eavesdropped by those with scanners. It also allowed relatively easy cloning of phones, allowing hackers to steal identities and then run up large bills. Towards the end of the 1980s these issues were becoming extremely serious.

- Another problem was fragmentation, with different technologies in use in many countries. This prevented economies of scale as well as any form of roaming.
- Cost of delivering services was high with newer technology promising savings.
- Finally, the quality and capacity of analogue transmissions was relatively low and evolving technology allowed better voice quality to be delivered using digital solutions.

2G resolved all of these problems. It added security that has not, even 30 years later, been breached in any material manner. It eventually provided a major improvement in voice quality through digital encoding[10], and through a mix of new spectrum and better reuse of frequencies it enabled a growth in capacity that led to a network able to support an order of magnitude more users. Fragmentation of technology persisted but to a lesser degree. Europe consolidated on GSM while the US and Japan had their own technologies. As a result, roaming was possible throughout Europe and to other countries in the world that had adopted the GSM standard but remained difficult elsewhere.

The use of digital technology also allowed 2G to offer data transmission, although this was seen at the time as of lesser importance. The unexpected and widespread use of the short message service (SMS or texting) showed the community the potential of data use, leading to subsequent evolutions such as GPRS (the General Packet Radio Service) enabling data capabilities better matched to requirements. This was a highly successful evolution, delivering a materially better technology that overcame the problems of the previous generation and set the stage for the dramatic growth in adoption of cellular technology.

As we have seen, the 1990s saw the increasing adoption of the Internet on desktop computers. Those involved in cellular technology predicted that the benefits of the Internet would be wanted while mobile but understood that 2G could not provide sufficiently high data rates to deliver an attractive service.

---

[10] Although some of the earlier voice codecs were relatively poor, these improved over time.

## A very abbreviated history of telecoms

There was also a view at the time that video calling would evolve from voice calling and that the 3G mobile network would need to support this. Hence, the key objectives of 3G were:

- To deliver higher data rates to enable Internet browsing.
- To provide support for video telephony and the use of cameras to add pictures to texts (the MMS service).
- To enable both of these through delivering greater spectrum efficiency allowing much higher data throughout.

The emergence of 3G (also known as UMTS – Universal Mobile Telecommunications Service) happened at a time when a new access method had been pioneered by Qualcomm in the US as one of the 2G solutions adopted there. Termed code division multiple access (CDMA) it promised significant improvements in spectrum efficiency through evenly spreading interference across all users. After much debate, it was decided to adopt CDMA as the underlying technology for 3G around the world.

However, effective implementation of the 3G version of CDMA proved difficult. Early 3G networks did not deliver high data rates and were difficult to plan and manage. Cell sizes in the frequency bands provided at 2GHz were small, resulting in the need for many new base stations. Cells "breathed" under loading, reducing in size as more customers accessed them. The mix of circuit-switched and packet-switched traffic proved difficult to manage. Perhaps this did not matter overly as mobile Internet adoption was slow, with users finding it very difficult to browse on the phones of the early 2000s with their modest screen sizes and internet content not developed to render in multiple formats. Also, video telephony did not prove as popular as envisaged, with small screens, ill-placed cameras and high per-minute costs.

Evolutions of 3G slowly addressed these problems, with high-speed packet access (HSPA) finally enabling the data rates originally promised, and even exceeding them. This coincided with the introduction of the iPhone with its easy-to-use large screen and user interface which transformed Internet browsing, causing extremely rapid growth in data demand.

Despite the improvements delivered by HSPA in its various forms, it was clear that 3G had not completely satisfied user performance demands. While various generations of HSPA radically improved the data rate, the latency of the technology (i.e. the delay till the first information arrived following a request) was still unacceptably long. The mix of circuit and packet switching made the networks less efficient and costlier to manage.

The aim of 4G was to fix these problems. It did so with a different air interface termed orthogonal frequency division multiplexing (OFDM) and by the removal of circuit switching. It also used wider frequency channels and importantly it significantly reduced the latency. The lack of circuit switching meant that voice calls could not be handled in the manner adopted in previous generations and only some five years after its introduction was voice finally carried over 4G using "voice over LTE" (VoLTE).

The more stable and data-optimised networks offered by 4G meant that higher data rates could effectively be delivered to mobile users. Hence, the perception of most was that 4G was significantly faster than 3G. It was also around 2.5 times more spectrum efficient than 3G allowing an important improvement in network capacity.

In summary there was a clear need for each generation up to and including 4G:

- 1G delivered mobile voice capability for the first time beyond a few professional user groups.
- 2G increased capacity for the rapidly growing demand as well as tackling clear problems with 1G such as security.
- 3G was designed on the premise that the Internet, widely used on fixed computers, would be valuable on a mobile device.
- 4G resolved the capacity and quality challenges that were found for 3G as well as delivering more capacity and somewhat higher rates.

Each generation was a solution to an identified problem. But as the manufacturers came to design 5G it was not obvious what problems it was resolving. 4G was working well. Instead of solving a clear problem, designers of 5G concluded that a key feature of previous generations had been faster data

## A very abbreviated history of telecoms

rates and hence this should also be the case for 5G. The case for this is shown in the chart below which sets out typical achieved mobile data rates versus date of introduction for each mobile generation (note that the y-axis is logarithmic).

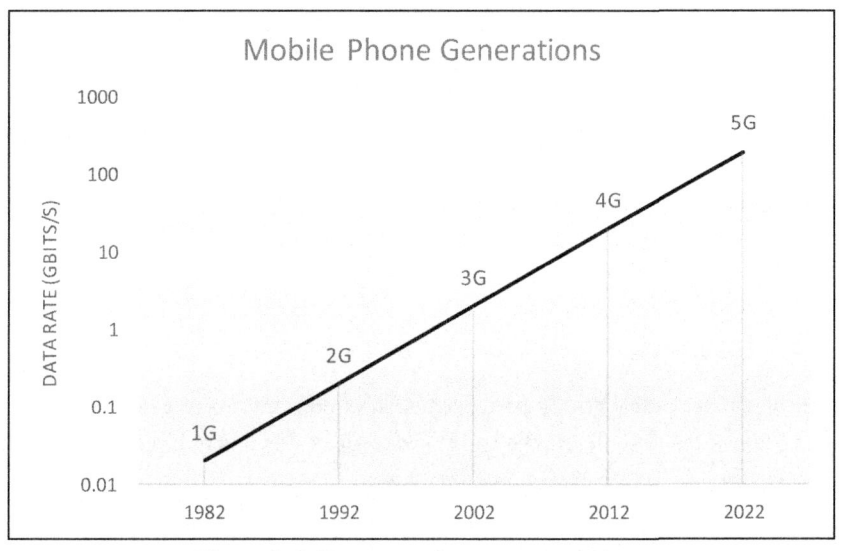

Figure 2-1: Data growth across generations

Two aspects are clear from the figure – a regularity in timing and a steady improvement in data rates.

Generations have appeared every decade with great consistency. This may be predominantly because all the various steps needed from research, through standardisation, to design and production take this long. It may also be somewhat self-fulfilling as the industry anticipates a cycle of this length and so tends to work towards it. In fact, 5G actually appeared around 2020 – slightly ahead of schedule, mostly for geopolitical reasons I will discuss later.

Each generation has also, approximately, resulted in a ten-fold increase in data rates. This is somewhat more difficult to see, as each generation has evolved during its decade, often improving its data rates throughout its 10-year period and so picking any one data rate for a particular generation is somewhat arbitrary. Also, the peak data rates quoted are rarely realised in practice. The chart above aims to select practical data rates. Hence the assumption of around

200kbits/s for the GPRS element of 2G, around 2Mbits/s for the HSPA evolution of 3G and around 20Mbits/s for the early deployments of 4G. Whether these are exactly right is of less relevance than the observation of the trend.

Just as with fixed, my thesis is that the race to deliver sufficient data rates had been run - reaching its conclusion with 4G. As I discussed in "The 5G Myth" and summarised earlier, with good 4G connection, mobile data rates are no longer the constraining factor on performance – instead it is the Internet servers or the screen size of the device. Hence, the extrapolation for 5G was a step too far. 5G became a solution in search of a problem and that problem is yet to emerge, and likely never will, as I discuss in detail later.

## 2.4 Broadcast

I cover broadcasting briefly here because it is both a form of communications and because, as we will see, the consumption of video content has a huge impact on the speeds and capacity of other networks.

Broadcasting started out with radio transmissions in the early 1920s. An increasing number of radio stations were broadcast, so much so that they started to interfere with each other, heralding an era of spectrum regulation. With cinemas evolving rapidly - showing the first synchronised sound and video film in 1927 - it was clear that there would be a huge demand for video broadcast. Technology was rapidly developed enabling the first video broadcasts in the 1930s, although it was not until the 1950s that most could afford a TV receiver.

The initial problems to resolve were picture quality and channel choice. Early solutions had insufficient lines and were black & white, but quality steadily improved into the 1950s and 1960s. The number of channels was restricted by the amount of radio spectrum available which could not easily be modified.

Two approaches were adopted to deliver more channels.

1. Use an alternative delivery mechanism. For many years this was the cable network. Very early deployments took place in the late 1940s but it was not until the 1980s that over half of US homes were able to

connect to cable, with similar timing in other countries. Cable enabled a much greater bandwidth, allowing many more TV channels to be broadcast.
2. Improve technology. A key step was replacing analogue TV broadcasts with digital. This increased capacity around 8-fold and also paved the way for high-definition transmissions.

By the late 2000s the issue of channel availability had broadly been solved, with most having access to sufficient channels, and the quality of reception was rapidly improving with high-definition and ever-better displays.

Another problem that had long been recognised was the need to watch a TV programme when it was broadcast – so called "linear TV". By the 1980s video cassette recorders (VCRs) delivered somewhat of a solution. Programmes could be recorded at home and watched at a later date. But VCRs were often difficult to use, required forethought to set up the recording, and required management of video tapes. Programmes could not be watched until after the recording had finished and the tape rewound.

Alongside VCRs emerged the digital video disk (DVD). DVDs differed in that they were generally view-only – while there were recordable video disks they were not widely used. DVDs spawned Netflix and paved the way for many of the commercial players we see today, if as a technology it did not really add much and was quite quickly supplanted by streaming.

The solution to the tyranny of the schedule, as well as removing any limit on the number of channels broadcast, was to move to streaming where each viewer had a separate video feed delivering the content that they wanted at the time that they wanted it. But streaming is not broadcasting – which sends the same signal to many. Instead, streaming required a dedicated connection to each viewer, delivered typically via the home broadband, but also via a mobile phone for those watching on their screen.

As with fixed and mobile, there have been many developments, technical and commercial, driven by clear need:

- The desire to receive sound and then video in the home.
- The clear need for greater video quality.
- A desire for more TV channels to choose from.
- The desire to watch content of choice at a time of choice.

There have also been some missteps on the way, including 3DTV, and the desire for higher quality beyond high-resolution is not clear. Some programmes are now made in 4k quality, but the demand for 8k appears very weak.

Conventional broadcasting persists over terrestrial networks (DTTV), satellite and cable networks and there are still many live events such as major sporting fixtures, that many want to watch simultaneously. But all of these broadcast channels are in decline with falling subscriber numbers and advertising revenues. There will come a point for each when maintaining the platform is uneconomic since for all platforms the costs are relatively fixed, so as subscriber numbers fall the cost per subscriber will rise. And many networks are already at the point where major equipment replacement is needed – new terrestrial transmitters or new satellites. Hence, despite the desire for some live broadcast viewing, eventually the platforms will be shut down. This may take many decades, especially where politicians intervene to keep key broadcast networks alive longer, but streaming is clearly the longer-term solution for video consumption and does allow simultaneous viewing of live events, albeit via multiple individual streams all sent at the same time. Audio broadcasting may last for longer as it is used in cars and elsewhere.

A shutdown, or reduction in scope, of the terrestrial broadcast network would free up some UHF spectrum in the bands 400-700MHz. Spectrum liberated from broadcasting has already been used to enable better mobile networks – the 800MHz band for 4G and the 700MHz band for 5G. Freeing up even more spectrum might provide some help in delivering better rural coverage as discussed in Chapter 6, although this spectrum is unlikely to be transformational given the large amount of sub 1GHz spectrum that the mobile operators already deploy.

With broadcasting the end of history is clearer than in fixed and mobile, as broadcasting switches to streaming. That places additional requirements on fixed and mobile networks – requirements which they are already well able to meet.

## 2.5 Shannon

In 1948 Claude Shannon published "A Mathematical Theory of Communication". This set out the fundamental limits on the data rates that could be delivered given the bandwidth and signal-to-noise ratio (SNR) in any communications channel. Since then, communication systems have been approaching these limits.

In fixed networks the use of fibre optic cables provides enormous bandwidth and hence the Shannon limit is rarely reached. More capacity can be provided by using additional wavelengths (more bandwidth). Only where there are copper cables is it an issue, and bringing fibre closer to the end point helps by shortening the copper connection which in turn improves its SNR.

The position is completely different in mobile networks. Here there is limited bandwidth and often low SNR. Many techniques have been adopted to reach close to the Shannon limit including advanced coding schemes and adaptive modulation. Mobile systems are now considered to be operating so close to the limit that improvements in the technical efficiency of each channel are no longer possible. In recent years there has been interest in finding more "channels". The Shannon limit applies to each communication path to the user. If communications could be sent via multiple paths then in principle more data can be transmitted. This can occur in mobile networks where radio signals take multiple paths to the end user, bouncing off various buildings and obstacles. MIMO antenna systems utilise these multiple paths to increase overall data capacity. However, the evidence from 5G deployments is that they can only provide limited gains of perhaps 50% above standard communications while maintaining reasonable processing and power consumption requirements.

Mobile systems can have near-infinite increases in capacity through using smaller cells. Each cell has a set capacity so the more cells, the greater the capacity. Also, the SNR tends to be higher in smaller cells, supporting greater data rates. But deploying large numbers of smaller cells is expensive and can only be justified if they result in increased revenue.

Broadly, fundamental limits mean that there are limited gains from new technology – and now that we have reached these limits the benefits of new generations of mobile systems are small – another clear reason why history is coming to an end.

## *2.6 Geopolitics*

An area of very active telecoms history, but as we will see an unhelpful one, is geopolitics.

For much of telecoms history geopolitics played a minimal role. Countries worked together to enhance technology and at the International Telecommunications Union (ITU) to deliver global regulations. Standards bodies that were formed at national and regional levels merged into global standards entities such as the 3GPP. Suppliers such as Ericsson became truly global, and many operators expanded across multiple countries.

While there have been ebbs and flows in the extent of multi-national operators over the years, a spirit of openness and embracing the best technology wherever it came from, pervaded much of telecoms history.

All that changed around 2020 when countries started to ban Huawei and ZTE equipment from being deployed in their country and some required existing equipment to be removed. A US ban followed in 2022. Instead of international cooperation the environment progressively became one of international competition. Politicians saw it as important that their country was at the forefront of deploying new technology such as fibre and 5G, and that ideally the technology was designed and manufactured locally to remove reliance on external suppliers.

Apart from distorting the equipment supply market, the main effect has been pressure in many countries to deploy technology for the sake of it – to be seen as being a global leader or being high up international league tables. This has led to many "Gbit" initiatives for broadband and many targets for 5G deployment. For example, the EU recently bemoaned[11] the slow progress against its (unjustified) targets for 5G deployment across the continent. To these politicians, since connectivity is a good thing then more of it must be better. Fibre must be superior to copper and 5G must be superior to 4G. And, superficially this makes sense – these technologies are better in so much as they deliver higher data rates. The problem, as I have already hinted at, and will discuss in much more detail in the next chapter, is that the "betterness" that is delivered is unnecessary. Hence, cost is being incurred for no good reason.

The geopolitical situation might, then, make it appear that history is far from ended. There is still much fibre to be laid and 5G, and then 6G, to be deployed. There are many local suppliers to be nurtured and emerging standards to take a lead in. But these are all misguided political reactions, and while politicians can do much to prolong inappropriate strategies, in the end the market will impose reality – as indeed it is in the EU as will be discussed in Chapter 6.

Politicians have learnt a little from history, but insufficient to understand that demands are eventually met, and that telecoms history is coming to an end. As I will show in Chapter 6 there is much for them to do in delivering ubiquity and this is where attention will turn as it steadily becomes clearer that it is what consumers value.

## 2.7 Becoming a utility

History occurs when there is change. Once everything has settled down then – at least as far as Fukuyama was concerned –history is over.

We can see that in telecoms. Technological breakthroughs are rare, and operators are becoming utilities.

Technology changed materially in fixed networks up to the 2000swith modems and with new forms of fibre, but there has not been material change since then.

---

[11] https://digital-strategy.ec.europa.eu/en/consultations/consultation-white-paper-how-master-europes-digital-infrastructure-needs

In mobile there was huge innovation with digital in 2G, CDMA in 3G, OFDM and MIMO in 4G but, despite all its claims[12], the 5G radio is a slightly more flexible version of the 4G one.

It is on the operator side that we see the least change. The market structure in which MNOs and fixed operators resides has been static for decades: typically, the same companies have been selling broadly the same thing for over 20 years, albeit with ever-increasing monthly data buckets on mobile networks (fixed networks tend to be unlimited). Indeed, they are progressively delivering fewer services as fixed operators turn off their copper networks[13] which turns off "native" voice call support, requiring customers to use OTT services such as WhatsApp. The greatest innovation has often been around additional benefits offered such as free theatre tickets.

Both fixed and mobile operators are rapidly tending towards "bit pipes" by which I mean their only role is to transport data from one place to another. The term "bit pipe" is often used pejoratively but it need not be. The bit pipe is an indispensable part of our communications world; indeed it is the most fundamental and critical element. Delivering a bit pipe is a large responsibility. It also ought to be a fabulous business – with every home paying every month for fixed broadband and every person over the age of about twelve years[14] paying every month for mobile connectivity. Income is guaranteed – connectivity is one of the very last things people will forego. Most businesses would love to have an utterly reliable monthly revenue stream coming from almost every person and house in the country. In later chapters I will look at why operators have broadly failed to capitalise on this, but in essence the problem is their misreading of "history". They still believe that there is growth, new technology and new applications that they can use to deliver increased revenue and they have invested accordingly.

---

[12] There is something very Orwellian about terming the 5G radio interface "New Radio" when it is just a tweak on 4G.
[13] See eg https://www.openreach.com/fibre-broadband/retiring-the-copper-network
[14] https://www.statista.com/statistics/1326211/children-owning-mobile-phone-by-age-uk/

# A very abbreviated history of telecoms

Operators, especially MNOs, have long tried to avoid their role as bit-pipe providers. The list of additional services they have tried and failed to monetise is depressingly long and includes video calling, push-to-talk, location-based services, picture messaging (MMS), walled garden Internet services, pay services[15], converged offerings, sports packages, IoT leadership and so much more. Even now they are trying to expand into other areas with private network provision where it seems likely that specialised suppliers of such networks will be more successful. Each time they fail badly but never learn the lesson. It is not that they are incompetent at implementing new services (although they generally are not particularly good either) but that they can never compete with a single provider with global reach. Take location-based services. If each MNO developed their own there would be around 500 efforts, whereas Google maps (and equivalents) can instantly be available on all networks with economies of scale that result in much better offerings. Each MNO developing their own WhatsApp equivalent now clearly looks daft, but many tried to do it.

Operators clearly are utilities – organisations that deliver critical resources that we cannot live without - and nothing more. We are used to utilities in electricity, water, sewerage and natural gas – telecoms has become another utility. The role of the fixed operators is to complete the deployment of fibre networks to the extent sensible and then just ensure those networks keep working. Similarly, the role of the MNOs is to keep their 4G and 5G networks operating reliably.

Just as with geopolitics, the end of history for organisations such as operators has not been universally recognised. This is primarily because the operators themselves do not see being a utility as an attractive outcome, even if their investors know it. But their preference is irrelevant. They are utilities and their "history" has come to an end. The only history left is the consolidation that will likely occur towards single networks – it is rare to have competing utilities.

---

[15] There are exceptions here in developing countries where MNOs have been successful at building a banking service.

## 2.8 Wrapping up history

This chapter has set out how, for much of history, telecoms in the form of fixed, mobile and broadcast networks, has addressed clear needs. Initially these were for basic voice or audio connectivity. Then, as the Internet intruded on our lives, the need for data grew. This led to fibre being deployed ever more deeply in fixed networks and data capabilities becoming the key design aims of 3G and 4G mobile solutions. For much of the Internet era – broadly the year 2000 onwards, networks were in a race to deliver sufficient data capabilities to meet demand. And when they did, demand itself often grew as new applications such as streaming video came about.

But the need for ever faster connections must eventually come to an end, and it has done so with the deployment of 4G and FTTC. After this time there was no clear problem for new generations of telecoms technology to solve. This led to those involved hoping that if they "built it" then demand would come. But as we will see in coming chapters this always looked overly optimistic.

In the next chapter I turn to looking in more detail at why user requirements, at least for those with good connectivity, have been met.

# 3 User requirements are nearly met

## 3.1 Introduction

In previous chapters I have hinted at how faster data rates are not required and mentioned in passing that while demand for data is still growing, that this growth is coming to an end. This chapter shows why both of these factors are happening.

Data rates and data capacity are separate matters, although there is some relationship in that if users demanded higher data rates, they would likely consume more data (eg by watching higher resolution video).

While there is increasing acceptance that higher data rates may not be needed, the understanding of what is happening to data demand growth is much weaker. The fact that demand has nearly plateaued is a key reason why telecoms history is at an end.

## 3.2 The Internet

I have noted how it is the Internet that drove much of telecoms history in the past half century, and in particular the need for ever faster speed. So how much speed do we need?

Demand for the highest speeds and the largest data volumes is almost invariably driven by video consumption. A person can only watch one video stream at a time, so understanding the data rates associated with the highest quality video feed required is a good current upper limit. By way of example, high definition requires 3Mbits/s and 4k video requires around 15Mbits/s[16]. When I wrote "The 5G Myth" in 2016 Netflix was recommending 5Mbits/s and 25Mbits/s, respectively so over the last eight years, peak speeds requirements have actually

---

[16] For example, see Netflix recommended data rates https://help.netflix.com/en/node/306. Here they suggest 3Mbits/s for HD and 15Mbits/s for 4k UHD.

*fallen* to around 60% of prior levels. Data rates for 8k video are much higher[17] but demand and supply of 8k is weak and for most the benefits over 4k are minimal. So, for a household where one person is watching 4k and two people are watching HD, the total speed required would be 21Mbits/s. That is just 5% of a gigabit connection. And it is more likely that this will decline rather than increase in future.

Mobile screens are far too small to make watching video at high resolution worthwhile. MNOs have found that "throttling" video to 1Mbits/s or even less has no noticeable impact on the viewing experience of those using mobile handset.

A somewhat different question is the speed needed for instantaneous web browsing. The issue here is less one of absolute speed and more of "latency" – the time taken for a request (eg for a new page) to be sent to a server and a response received as shown in Figure 3-1. Beyond a certain speed, other factors such as the maximum turn-around time at the server and the delays inherent in the Internet TCP/IP protocols become constraining[18] [5]. This data rate is currently around 10Mbits/s (and hence most users will not notice an improved browsing experience once data rates rise above this point). Resolving this requires changes to Internet protocols and architectures – something that has to occur on an international basis within Internet standards bodies and key industrial players.

This data is widely replicated. For example, Cloudfare[19] assessed the situation in detail and noted that:

> In the 2010 paper from Google, the author simulated loading web pages while varying the throughput and latency of the connection. The finding was that above about 5 Mbps, the page doesn't load much faster. Increasing bandwidth from 1 Mbps to 2 Mbps is almost a 40 percent improvement in

---

[17] Although there is some current thinking that AI might be able to generate an 8k video feed locally from a 4k video stream and hence there would be no need for higher bandwidth.

[18] For a detailed discussion see https://gettys.wordpress.com/2013/07/10/low-latency-requires-smart-queuing-traditional-aqm-is-not-enough/ .

[19] https://blog.cloudflare.com/making-home-internet-faster

User requirements are nearly met

page load time. From 5 Mbps to 6 Mbps is less than a 5 percent improvement.

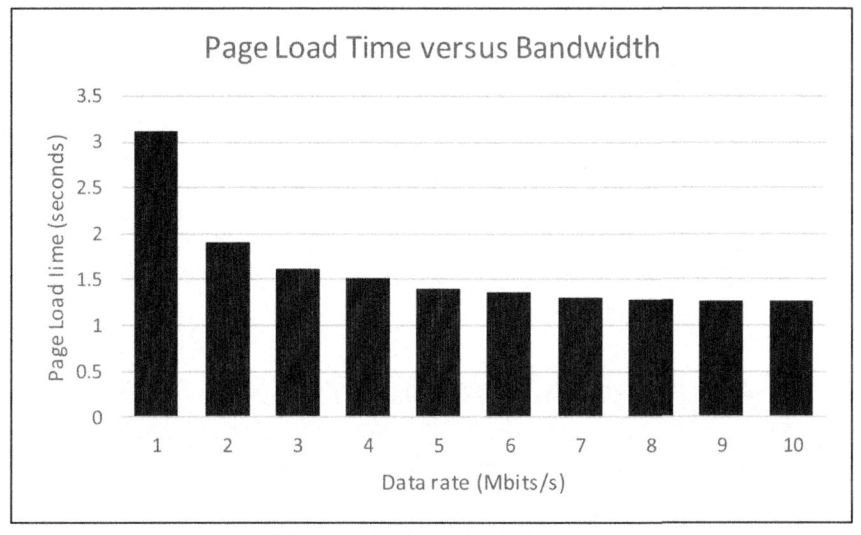

Figure 3-1 – Page load times

A more recent paper, published in 2023[20], entitled "Understanding the Metrics of Internet Broadband Access: How Much Is Enough?" assessed the rate needed for *fixed* access and came to similar conclusions albeit at a higher data rate of around 20Mbits/s. It states:

> Above about 20 Mb/s, adding more speed does not improve the load time. The limit on the load time is the latency to the servers providing the elements of the web page.

It also makes the point that this speed is increasing somewhat – from 13Mbits/s in 2013 to 20Mbits/s now, and postulates that this is because web servers are being upgraded.

---

[20] https://papers.ssrn.com/sol3/papers.cfm?abstract_id=4178804

Understandably, there is no clear single answer and rates will vary over time. Video rates may fall, whereas the rate at which there will be no further improvement in Internet experience may rise. Perhaps by 2035 the useful upper rate might have hit 35Mbits/s, but other factors may limit a further rise such as inherent latency.

Overall, there is a clear conclusion. Beyond a certain speed there is no benefit in going faster. The user experience will not change. This speed appears to be around 10Mbits/s for mobile connections and around 20Mbits/s for fixed connections. As noted, there may be multiple fixed users in a house, so 50Mbits/s per house may be a safer upper limit. This will align with the experience of most who have paid for packages of speeds beyond this and not noticed any improvement as a result.

To emphasise the point – 4G can easily deliver more speed than is needed as can FTTC. There is no benefit in going beyond this. Nor is one likely to emerge unless we dramatically re-engineer the Internet.

Many countries have gigabit targets – the desire to deliver data rates of 1Gbits/s or more via broadband. As the conclusion above makes clear, this is pointless. There is no benefit above around 50Mbits/s and requiring 1Gbits/s could result in significantly more expensive solutions, especially where fibre is difficult to deliver and alternatives such as FWA would suffice. A gigabit target could also result in consumers waiting longer for a solution – they could have, say, 100Mbits/s in a year or 1Gbits/s in 5 years. **Gigabit targets are not appropriate.** (But, as noted earlier, replacing copper by fibre has other benefits such as lower operational cost and greater reliability and is worth pursuing on that basis.)

Having dealt with speed, I now turn to capacity where I concentrate on mobile capacity. This is because fixed capacity is, in essence, unlimited. Most broadband subscribers have a dedicated connection and the amount they use it makes little difference, other than requiring relatively low-cost core network upgrades. But matters are completely different on mobile networks where capacity is shared and overall usage drives investment. Having said that, the same factors are driving growth on mobile and fixed networks and as we will see the same trends in data growth rates apply to both.

## 3.3 The iPhone effect

In the early 2000s mobile data was much sought but not found. 3G had been deployed on the assumption that mobile data demands would be huge, and yet virtually none materialised. That all changed in 2007 with the launch of the iPhone. Figure 3-2 shows what happened to both voice and data volumes in the four years immediately after the launch of the iPhone.

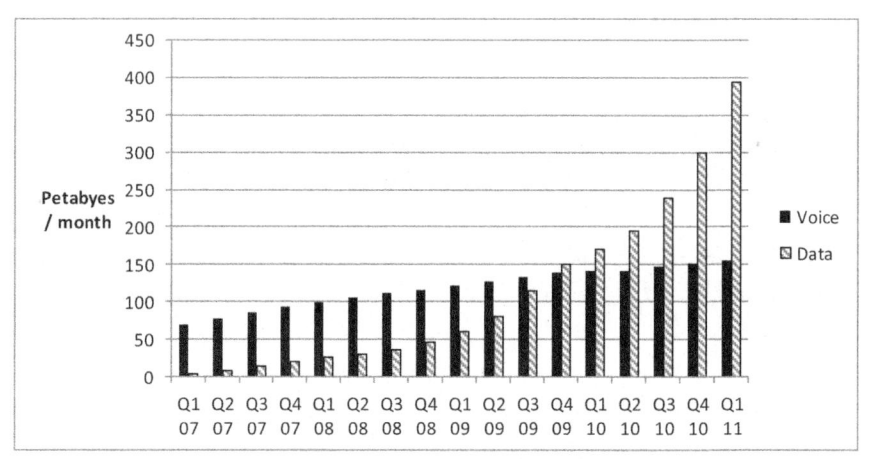

Figure 3-2 – Growth in mobile data in years after the iPhone was launched

While voice grew modestly, data demand exploded, growing around 100-fold in the five years from 2007 to 2011. It has continued to grow rapidly since. However, even in 2016, it was clear that growth was slowing. This is shown in Figure 3-3 where actual growth to 2016 and the growth then predicted to 2020 is shown along with a linear trendline.

The end of telecoms history

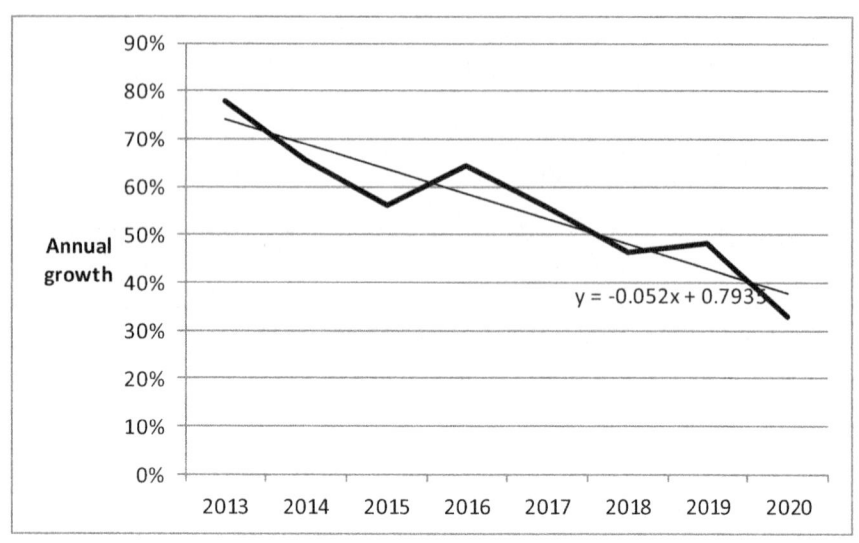

Figure 3-3 – Growth rates and regression line predicted in 2016: Source Webb, "The 5G Myth"

I predicted back then that the mobile data requirements over the period 2007 to 2027 would be as shown in Figure 3-4.

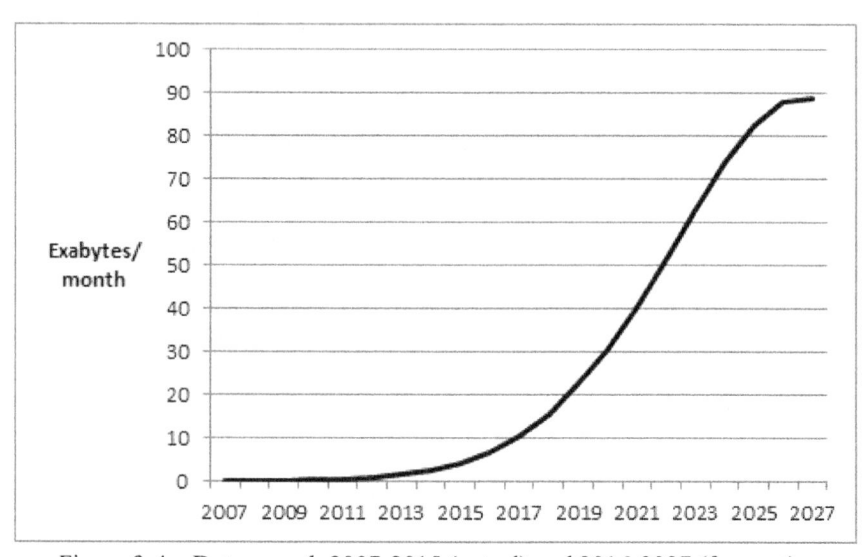

Figure 3-4 – Data growth 2007-2015 (actual) and 2016-2027 (forecast)

This shows a classic "S-curve", as experienced frequently with many new services and devices. I predicted that data would plateau at around 15-20Gbytes/user/month. Such a plateau might be expected. There is only so much data that a mobile subscriber can consume. Once they are watching video for all their free moments while downloading updates and attachments there is little more that they could usefully consume. At the level of the plateau shown here the average mobile user would be consuming over an hour of video via cellular on their mobile device every day. (Note that there may also be video streaming to the mobile device via Wi-Fi, and video streaming to other devices such as tablets also likely via Wi-Fi so this represents video mostly consumed on the move.) Of course, some users will exceed this, but equally many will consume much less or will download in advance.

As I will show shortly, this forecast is looking accurate. Actual data to 2023 fits the curve well. Before turning to the numbers, I want to look at why video traffic is the only traffic we need to be concerned about.

## 3.4 Video drives usage

It has long been reported that mobile data consumption is dominated by video. For example, in 2022 Ericsson reported[21] the data in Figure 3-5 (I will return to their conclusion of a "surge in mobile traffic" later).

It is clear from the chart that video dominates – consuming in the region of 70% in 2022 and an anticipated 80% or more in 2027. The next largest category – social media – is tending more towards video itself so might be included in the video category.

A few simple calculations make it clear why this is. An hour of HD video equates to 1.3GBytes of data. Even when throttled to 1Mbits/s this is still over 400Mbyte of data. By comparison browsing the Internet uses about 15Mbytes/hour (depending on the content and speed of clicking) – less than 4%. At present, there is no content that comes close to video.

---

[21] https://www.ericsson.com/en/reports-and-papers/mobility-report/reports

## Video Drives Surge in Mobile Data Traffic

Estimated global mobile data traffic by application category (in exabytes per month)*

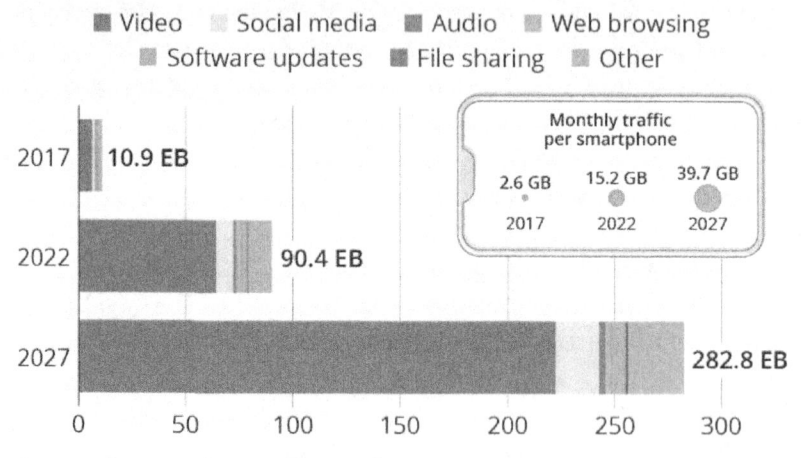

- Video
- Social media
- Audio
- Web browsing
- Software updates
- File sharing
- Other

2017: 10.9 EB
2022: 90.4 EB
2027: 282.8 EB

Monthly traffic per smartphone:
- 2017: 2.6 GB
- 2022: 15.2 GB
- 2027: 39.7 GB

\* one exabyte equals one million terabytes
Source: Ericsson Mobility Report

Figure 3-5 : Video consumption

The implications of this, partly as noted earlier are:

1. There is an upper limit on mobile data usage since there is only so much video we can realistically watch.
2. When considering whether other applications might emerge that would materially change usage, the key criteria is their data rate compared to video.

These conclusions are also true for home networks where consumption is likewise dominated by video consumption. Here video rates can be higher as HD is standard and 4k occasional.

User requirements are nearly met

## 3.5 Data confirms predictions

### 3.5.1 Mobile data growth figures

Since 2016 the data does, broadly seem to have followed the S-curve discussed earlier.

For my main analysis I have taken Barclays Research data for Europe which reported growth in 2023 of around 23% as shown below. I have chosen this data because:

- It is from a very reputable source.
- By covering multiple countries, it averages across many different environments providing more useful input.
- It focuses on developed countries which tend to lead developing countries in their usage.

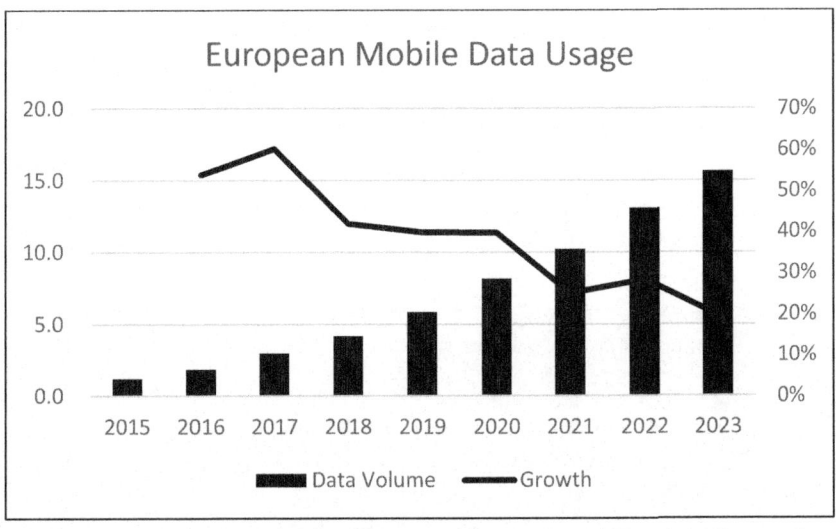

Figure 3-6 : European mobile data growth rates: Source Barclays Research

This shows a clear downwards trend, aligned with the S-curve.

Charting the Barclays data on the 2016 data and forecast from "The 5G Myth" gives the following chart.

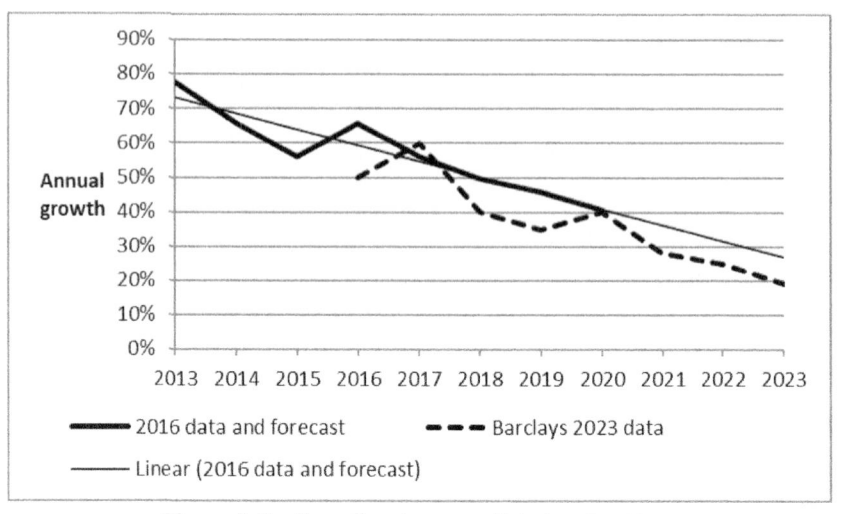

Figure 3-7 : Growth rates – predicted and outturn

From this it is clear that:

1. The growth levels from 2016 onward tracked quite well those predicted with the dashed line sitting close to the solid line.
2. For the 10 years since 2013 there has been a very clear and inexorable downward trend in data growth closely tracking a straight line and falling by around 5%-6% per year.
3. Data growth is falling faster than I predicted with growth falling away faster after around 2020.
4. There is no "5G effect". The introduction of 5G has not led to more mobile usage than would have been predicted.

This aligns near-perfectly with the 2016 S-curve prediction suggesting that the underlying behaviour will indeed follow the S-curve and will fall to near zero by around 2027, with significant variations by country.

There are many other data sources all pointing in the same direction. For example, Analysys Mason published[22] the following:

---

[22] https://www.analysysmason.com/research/content/articles/cellular-data-traffic-rdnt0/

User requirements are nearly met

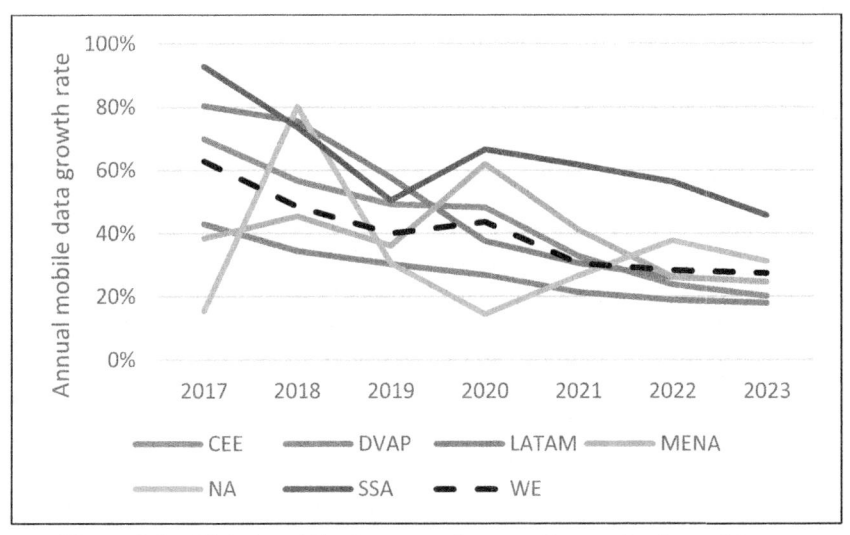

Figure 3-8 : Global mobile data growth rates: Source Analysys Mason

Although this is far from clear, especially in black and white, it shows that Western Europe (shown as the dashed line) is very much typical of much of the rest of the world and hence a good dataset to use. Only sub-Saharan Africa stands out as being materially different.

To emphasise, there is huge variation from country-to-country and data usage can depart from trend when, for example, new tariffs are introduced, but the evidence seems clear that growth rates fit usage data following an S-curve and will reach zero in the next five years. Indeed, it is striking that all of the curves shown above from multiple data sources all have the same underlying straight line downward trend.

To recap, there is a clear underlying theory as to why growth will slow based on a limit of personal video consumption and the well-understood mechanisms of consumer adoption leading to an S-curve. And there is over a decade of data that fits the model very well. The remainder of the forecast to the point of a plateau by around 2027 looks highly plausible on that basis.

Now to return to Ericsson's claim that there will be a "surge in mobile data growth". Firstly, to note that Ericsson have a strong self-interest in predicting strong growth since it leads to greater sales of equipment as well as making their solutions appear more important. Next, the chart shown above shows data growth of 9x between 2017 and 2022 but only 3x between 2022 and 2027. It is true that absolute volumes of data used have grown more in the latter period but growth rates have fallen by a factor of three – hardly a "surge". Finally, of course, they are making a forecast (just as I am). They, and others like Nokia, tend to accept that mobile data growth rates have fallen (since this is an incontrovertible fact) but do not openly state this. Indeed, Nokia predicts[23] growth rates might rise and postulates applications that might drive this – which I will spend the next chapter looking at. By way of a summary, Ericsson and many others who publish forecasts are not disinterested independent entities and their forecasts need to be treated accordingly.

Figure 3-9 shows a simplification of how some manufacturers and some regulators are predicting data growth.

None can deny that data growth is slowing – the data is incontrovertible. My view is that the clear trend of the last 10 years of data growth falling by 5%-6% a year on average will continue absent any major change in behaviour as shown by the straight downward line on the chart until it reaches near-zero[24]. It is a powerful trend and one that will take a lot to change. It is also the most logical approach – if data growth has been falling consistently for 10 years why would it not continue to fall?

---

[23] https://www.nokia.com/about-us/news/releases/2023/10/31/nokia-technology-strategy-2030-emerging-technology-trends-and-their-impact-on-networks/

[24] It is possible, but unlikely, that data usage could actually fall (ie growth becomes negative) if video compression continues to improve. Since the main use of data is video, if the data rates needed for a video stream fall, then overall data usage would fall too. However, this would not be particularly consequential. Data usage should also track population growth so may increase (or decrease) slightly as the population size changes.

User requirements are nearly met

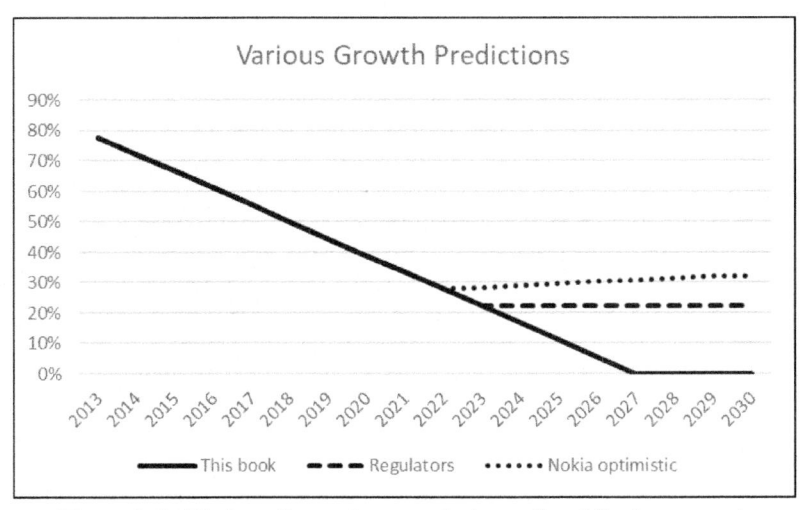

Figure 3-9 : Various forward extrapolations of mobile data growth

Regulators tend to have been caught out. Just two years ago Ofcom forecast[25] data growth in the region 25%-55% into the future. They are disinclined to drop this completely and now they and other regulators tend towards an assumption that data growth will level out at the current 20%. As the chart shows this looks strange – why suddenly flatline? And in Ofcom's case it was undermined by their report[26] that 2023 data growth was just 15%. Growth clearly continues to fall.

Manufacturers, who as explained above have a self-interest in seeing growth, can take a more extreme position, although to be fair Ericsson's most recent forecast[27] is more reasonable. The Ericsson predictions are shown below, for all regions firstly to see the overall trends, and then for just Western Europe and North America as the two regions that tend to be ahead of others.

---

[25] https://www.ofcom.org.uk/__data/assets/pdf_file/0036/248769/conclusions-mobile-spectrum-demand-and-markets.pdf
[26] https://www.ofcom.org.uk/research-and-data/telecoms-research/data-updates/telecommunications-market-data-update-q4-2023
[27] https://www.ericsson.com/en/reports-and-papers/mobility-report/dataforecasts/mobile-traffic-forecast

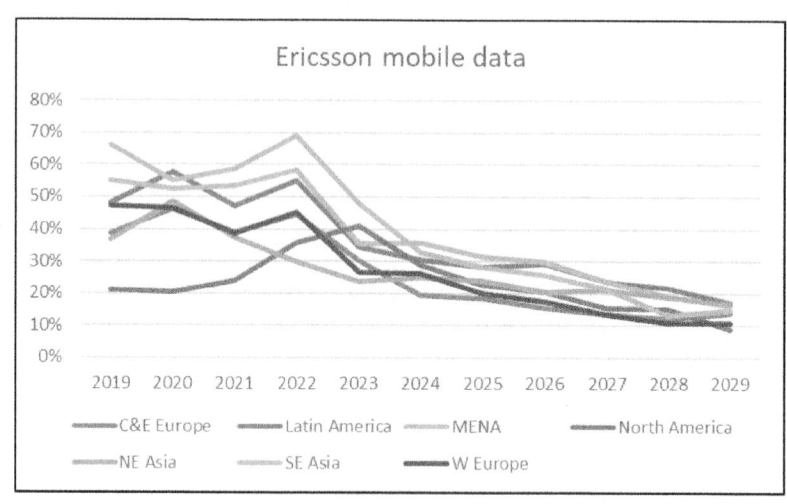

Figure 3-10 : Ericsson prediction for all regions

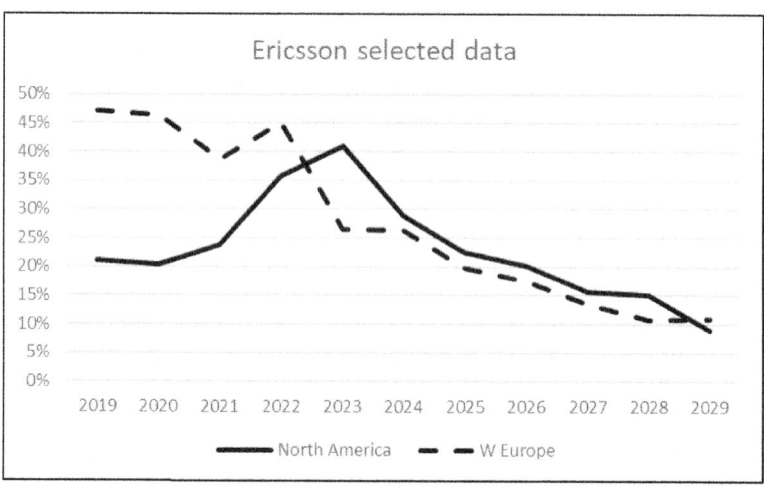

Figure 3-11: Ericsson prediction – W Europe and N America only

Ericsson, like me, are clearly predicting a slow down with trend lines that suggest an eventual plateau. The only difference is the timing. The Barclays data suggests Ericsson data is 5-10% too high and hence overly optimistic. Regardless, even Ericsson are predicting the end to data growth.

User requirements are nearly met

Conversely, Nokia[28] say:

> In the Global Network Traffic 2030 report, Nokia projects that end-user data traffic demand will increase at a compounded annual growth rate (CAGR) of 22% to 25% from 2022 through 2030. Global network traffic demand is expected to reach between 2,443 to 3,109 exabytes (EB) per month in 2030. If there is a higher adoption rate of cloud gaming and XR in the second half of this decade, Nokia projects a CAGR that reaches as high as 32%. Technology breakthroughs like XR and digital twins, combined with Web3 and other much-lauded emerging innovations, will transform businesses, society and the world.

Manufacturers cite technologies such as XR and FWA as drivers for growth and these are discussed in more detail in the next chapter.

The bottom line is that Figure 3-9 shows the ridiculousness of the positions of those who expect growth trends to suddenly change.

The implications of data usage reaching a plateau in the next few years are huge and I spend the whole of Chapter 5 looking at these. They include the fact that once we reach the plateau there will be no need for more spectrum (and so no auctions), no need for 6G, no need for additional cell sites, no need for capacity expansion of networks and little chance of increasing end-user prices. (In the next few years, before we reach the plateau it is possible some additional spectrum would be helpful.)

### 3.5.2 Fixed data growth figures

Fixed data requirements are also following the same curve, albeit with a significant "blip" around the Covid era – indeed growth rates are lower and heading into single figures. Figure 3-12 shows the Barclays Research data for Western Europe.

---

[28] ibid

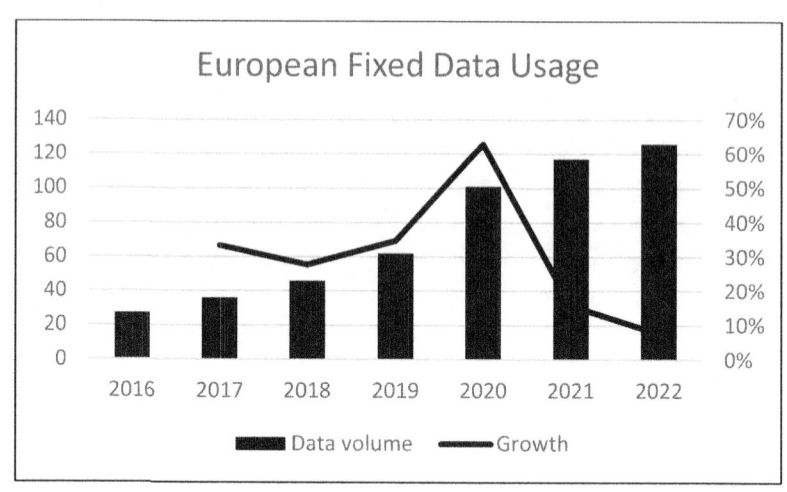

Figure 3-12  Fixed broadband growth rates: Source Barclays Research

Communication Chambers[29] provided a detailed assessment of global fixed line traffic growth[30] as part of their annual assessment. The global data is shown first.

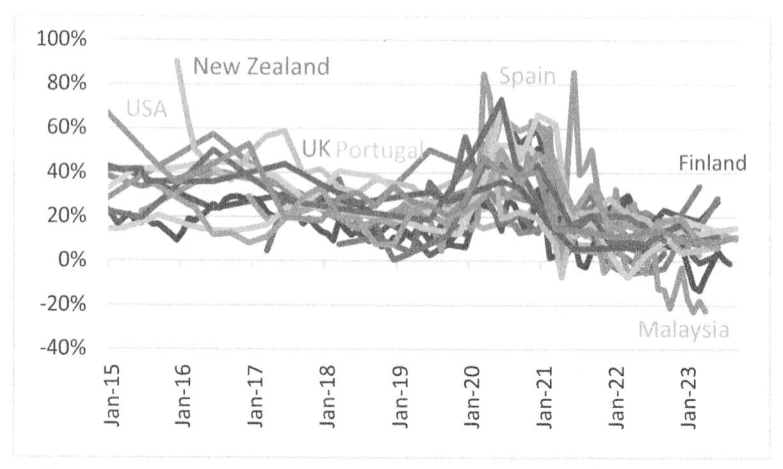

Figure 3-13: Global fixed line growth rates: Source Communications Chambers

---

[29] http://www.commcham.com/
[30]
https://static1.1.sqspcdn.com/static/f/1321365/28627748/1702742750033/Internet+Traffic+2023+v1.pdf?token=MbOXA%2F02GpoeElGYQKE69zKVlFg%3D

User requirements are nearly met

While there are a lot of lines the downward trend is clear as is the fact that all countries are experiencing growth rates below 20%/year. Taking out some of the lines by selecting English-speaking countries, gives the following chart.

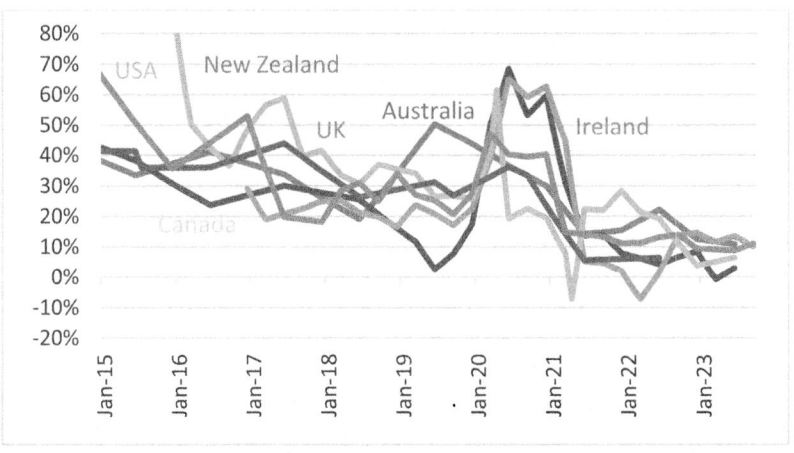

Figure 3-14: Global fixed line growth rates in English-speaking countries: Source Communications Chambers

The Covid19 spike is now very obvious, as is the trend towards 10%/year growth or less. Finally, the average across all countries is shown below.

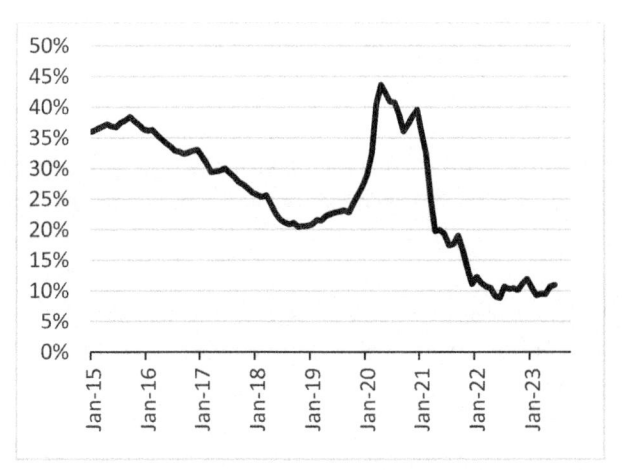

Figure 3-15: Average global fixed line growth rates: Source Communications Chambers

All the data clearly point to fixed line growth now being 10%/year or even less and falling.

## 3.6 Latency is sufficient, faster is prohibitively expensive

There is one further performance parameter that I have not mentioned much so far – latency. This is the delay between sending a request (eg for a new webpage) and getting a response. Fixed networks tend to have a latency below 10ms even where access is via Wi-Fi, whereas good 4G networks have latency in the region 30-40ms.

Latency is clearly important. If latency grows too long, then everything slows down and eventually becomes highly frustrating to use.

One of the key promises of 5G was "ultra-low latency" with the intent of dropping below 10ms. This, it was suggested, would allow real-time and tactile communications such as the remote control of vehicles.

Perhaps the most obvious point to make here is that if 5G achieves its latency goals (which it shows little sign of doing at present) then it will bring it into line with fixed networks. If lower latency were important then we might expect to see a number of applications benefiting from it already implanted using fixed networks and Wi-Fi. We have not seen such applications, at least not at any significant scale.

Many of the applications that 5G proponents suggested need low latency generally work well with 4G latency levels. For example, the advice for gamers is[31],

> Latency is measured in milliseconds and indicates the quality of your connection within your network. Anything at 100ms or less is considered acceptable for gaming.

---

[31] https://www.screenbeam.com/wifihelp/wifibooster/how-to-reduce-latency-or-lag-in-gaming-2/

User requirements are nearly met

As I set out in "The 5G Myth", delivering latency levels equivalent to those of fixed networks across a mobile network is challenging and expensive, both in the equipment needed and the resources consumed[32]. This is likely why few 5G networks have implemented the low latency features available within the standalone solution element of 5G.

## 3.7 We have all we need

This chapter has shown that for the first time in the 150-year history of telecoms, we finally have all that we need. Data rates beyond 10Mbits/s on mobile and around 50Mbits/s on broadband will not make any meaningful difference to end users – and we have these rates where we are well connected. We are consuming almost all the video we want and as a result data volumes are reaching a plateau. Calls for 5G everywhere and gigabit connectivity are misplaced and will result in capital expenditure on networks for no user benefit. We do not need, nor increasingly are willing to pay for, better networks.

Those with vested interest in seeing increased demand for speed and capacity suggest that new applications that will change this are just around the corner. The next chapter looks at whether this is likely.

---

[32] Low latencies tend to require networks operating at lower utilisation levels such that there is spare resource readily available when needed.

# 4 Is there anything that could reignite growth?

## 4.1 Moving demand requires something huge

In the peak years of mobile data growth, annual growth rates of 80% were reached. Regulators and others have been suggesting growth rates might be 20-40% a year going forward. I am suggesting growth rates will fall from 20% now towards zero.

To make a material difference – one that might require more spectrum, more cells or some other major change – would require at least a doubling of data consumption. In most networks individual consumption is around 10-15GByte/month. There would need to be an application that generated about the same again – another 10-15GByte/month per person on the network to make such a difference. If it were video-based it would need to generate about an hour of video per day (that was delivered via cellular, not Wi-Fi).

Even if such an application did appear, once it was adopted growth would return back towards zero. A new application of this sort would need to appear every three years or so to keep growth rates at the levels they were at when 5G was introduced.

As we will see, there are very few applications that can deliver at this scale.

## 4.2 The 5G hopefuls

We have been here before. To justify 5G, proponents suggested a plethora of applications that would be adopted, driving demand and delivering increased revenue. Back in 2016 the favourite candidates were:

- High-resolution video for person-person calls.
- The smart office.
- HD video sharing at venues.

## Is there anything that could reignite growth?

- Train passenger use of high-speed internet.
- Remote computing.
- Smart wearables.
- Sensor networks.
- Mobile video surveillance on transport.
- Tactile internet for remote control.
- Emergency service communications.
- Automated driving.
- Collaborative robots.
- eHealth monitoring.
- Remote surgery.
- Broadcast-like services.

Eight years later and few of these have emerged, none at the scale to make a difference. That is despite the Covid lockdown providing an environment where remote working and similar thrived. Back in 2016 I set out why I did not believe that these applications would make a material difference to network loading and the arguments I used then have broadly been proven.

Some of these applications have appeared, or grown in use – for example smart wearables, sensor networks and person-to-person video calls. But these uses still only account for a small proportion of overall data usage, and this is unlikely to change.

Some 5G proponents would counter that sufficient time has not yet elapsed for their favoured applications to appear and that the 5G network features that are needed, which include standalone (SA) network cores, are not yet widely implemented. And indeed, it was not until seven years into 3G deployment that the iPhone appeared and finally brought about the vision of data usage. We are still only about four years into 5G deployment so it may be too early to tell. Nevertheless, we ought to be more sceptical now towards claims that great new applications will emerge than were around in 2016.

A different approach is to ask whether, if they did emerge, these applications would result in the level of data traffic needed to generate a return to 30%/year data growth. The table below sets out the likely usage.

The end of telecoms history

| Application | Video? | Hours/day/person |
|---|---|---|
| High-resolution video for person-person calls. | Yes | 0.5 (some WiFi) |
| The smart office. | Yes | 3 (but WiFi and/or relatively low resolution) |
| HD video sharing at venues. | Yes | <0.1 |
| Train passenger use of high-speed internet. | Yes | 0.1 (via train network) |
| Remote computing. | No | |
| Smart wearables. | No | |
| Sensor networks. | No | |
| Mobile video surveillance on transport. | Yes | <0.1 |
| Tactile internet for remote control. | Yes | <0.1 |
| Emergency service communications. | Yes | <0.01 |
| Automated driving. | Yes | Unclear, likely low as cars are autonomous |
| Collaborative robots. | No | |
| eHealth monitoring. | No | |
| Remote surgery. | Yes | <0.01 |
| Broadcast-like services. | Yes | <0.1 |

Table 4-1 : 5G suggested applications

The first two – video calls and the smart office – could have the potential to add material amounts of video traffic. But these applications have broadly already happened, with WhatsApp style video calls and Teams/Zoom style meetings. Much of this traffic is handled over Wi-Fi and fixed networks and it has not materially shifted mobile network usage, apart from the growth "blip" in 2020 associated with the Covid emergency and seen in the data of the previous chapter.

Other applications, when averaged across the entire population, add little to overall traffic levels. For example, video surveillance might generate many hours of video per day but there are relatively few video cameras per person

Is there anything that could reignite growth?

(which are connected wirelessly). Take trains as an example. There are around 16,000 train carriages in the UK. If each carriage had a security camera, then that would amount to 1 camera per 4,500 people in the general population. Even if each camera generated traffic 24 hours a day that equates to just 19 seconds of video per person per day. Of course, there are other wirelessly connected applications as well, but the point is clear.

The set of applications put forward to justify 6G looks rather similar. For example, Ericsson says[33]:

> Examples of important 6G use cases include e-health for all, precision health care, smart agriculture, earth monitor, digital twins, cobots and robot navigation. These use cases can be sorted into three broad use case scenarios: the Internet of Senses, connected intelligent machines, and a connected sustainable world.

The EU set out[34] their understanding of the 6G vision as:

> The next generation mobile system, 6G, is described as a distributed intelligent network (underpinned by AI and machine learning), which creates 'interactions between the physical world, digital world, and biological (human) world, especially emphasising the real-time integration of cyber and physical spaces'

> It is considered that 6G will improve applications of previous mobile generations and introduce new ones, such as truly immersive extended reality (XR), high-fidelity mobile hologram and digital twins of real-world objects. Combining virtual reality (VR), augmented reality (AR) and mixed reality (MR), XR might have potentially interesting impacts in medicine (e.g. providing care outside hospitals), entertainment (e.g. online gaming), education and manufacturing industries (e.g.

---

[33]
https://www.ericsson.com/en/6g#:~:text=Examples%20of%20important%206G%20use,and%20a%20connected%20sustainable%20world.

[34]
https://www.europarl.europa.eu/RegData/etudes/BRIE/2024/757633/EPRS_BRI(2024)757633_EN.pdf

experimenting new designs without creating prototypes from scratch). According to an industry report, the current user-experienced data rate for 5G is not sufficient for XR seamless streaming to flourish.

Hologram technology will allow people to take part in a teleconference by representing their gestures or facial expressions and allowing remote technicians to perform remote troubleshooting or repairs. However, one study reports that 4G and 5G data rates may not enable such a holographic future because 'holographic images will need transmission from multiple viewpoints to account for variation in tilts, angles, and observer positions relative to the hologram'. Nevertheless, 6G might indeed allow the technology to succeed.

Advanced sensors, AI and high-speed connectivity ensuring low latency (e.g. the time data takes to transfer across the network), would allow replication of physical entities (including people, devices, objects, systems, and even places) in a virtual world. This digital replica of a physical entity is called a digital twin. Among the various use cases, digital twins might be used in medicine to build human immune systems, or in factories to simulate the deployment of specific complex deployed assets such as jet engines and large mining trucks, to 'monitor and evaluate wear and tear and specific kinds of stress as the asset is used in the field'. One think tank stresses how 6G performance (e.g. speed and latency) will better enable digital twin technology by providing real-time monitoring and control of physical objects and environments.

While it is not clear what some of these applications might entail, there is little here that differs from the 5G visions. The message seems to be that these applications did not emerge in 5G because it was not good enough, so 6G needs to be even better.

Is there anything that could reignite growth?

At a recent 6G conference[35], both Nokia and Ericsson suggested significant data growth driven predominantly by XR (the combination of VR and AR) and fixed wireless access (FWA).

I take a deeper look into those applications most stridently suggested below, covering:

- XR.
- AI.
- IoT and digital twins.
- FWA.
- Autonomous cars

My aim is not so much to predict whether they will emerge as whether they will generate new demands on networks or materially increase data usage. Is there any real "problem" here for which we need a better connectivity solution?

## 4.3 AR/VR

Go back a decade and XR (it was called VR/AR then) was postulated as one of the key reasons for 5G. What has happened over the last decade? VR certainly had every opportunity to thrive, with Meta pushing hard on the concepts of the metaverse, resulting in huge interest and many companies investigating what role the metaverse might play in their operations. Did it thrive?

---

[35] https://global6gsummit.com/

**Consumer VR headset sales by category, global, 2018-2028**

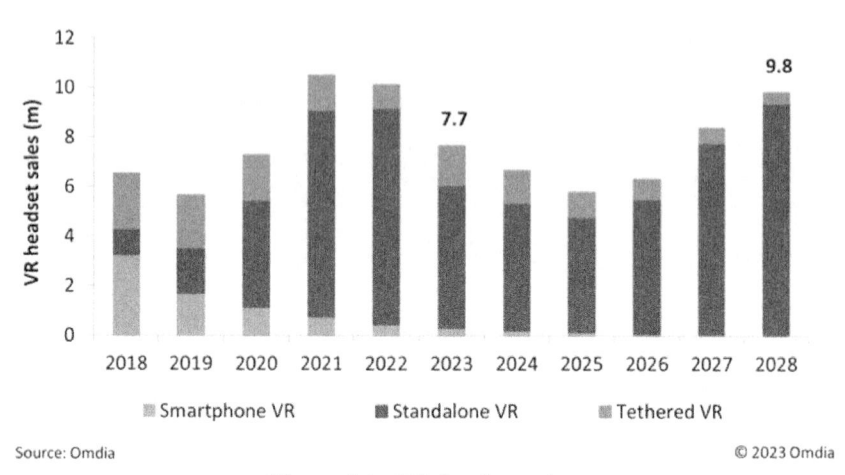

Figure 4-1 : VR headset sales

Figure 4-1 shows that VR headset sales peaked at around 10m/year in 2021. To put this into perspective the global handset market is about 1.1bn. VR headsets sales are less than 1% of handset sales. Since 2021 sales have been falling and are predicted to drop to 6m by 2025. Quite why Omdia think they will grow again after that is very unclear, but even on their numbers they never reach 1% of handset sales. The Apple VisionPro was supposed to change everything but there are many reports that sales have been disappointing[36].

Even more telling, none of these headsets have a cellular connection. They all use Wi-Fi to link to routers which are mostly connected via fixed broadband. So VR will, very clearly, not drive traffic on cellular networks.

What about AR? There has been much interest recently in AR "smart" glasses, and many models have emerged. Sales data does not seem to be readily available yet. A first point is that AR typically does not require much data. AR glasses

---

[36] Techradar suugests Apple have dramatically cut sales forecasts ( https://www.techradar.com/computing/report-claims-vision-pro-sales-have-fallen-sharply-as-apple-struggles-to-whip-up-demand ) while Quartz says they have shelved a new version of the headset ( https://qz.com/apple-stops-work-new-model-of-vision-headset-1851546812 )

Is there anything that could reignite growth?

superimpose images on top of reality. These images may just be directional arrows or text information, needing almost no bandwidth. Many of the "smart glasses" do not even do that but provide the ability to take photos and video and provide audio feed. This allows, for example, simple recording and upload of a video blog. This might amount to a few minutes a day of video upload – a level of content hardly noticeable on mobile networks. Other models of AR glasses can double as video screens, but typically then replace watching video on a handset, so do not change the overall traffic volume (indeed, they typically download the video to the handset then use Bluetooth to send it to the headset). So far it seems (1) that smart headset sales will not come remotely close to handset sales and (2) the traffic they generate will be small and may just replace traffic that would have been generated by the cellphone.

Those who claim XR is the killer application need to explain where all this traffic – and the revenue streams to pay for it – will be coming from. That they have not done this, at all, suggests that they are grasping at straws and repeating all the same mistakes as ten years ago. Less extended reality, more ignoring reality.

In summary, VR headset sales are minimal and all work using Wi-Fi. AR headset sales may be larger, but traffic levels generated equate only at most to a few minutes of video a day to those wearing them, which is likely to be less than 10% of the population.

It is very hard to imagine that the XR vision of Ericsson and the EU will come to fruition. But even if they do, the applications suggested will very likely use Wi-Fi as their local link.

## *4.4 AI*

There are two questions associated with AI and telecoms:

1. Will AI lead to increased traffic across telecoms networks?
2. Will AI enhance the capabilities of networks?

The first is most pertinent to this section but for completeness I will also address the second.

It does not seem likely that AI will materially change network traffic, at least across access networks. There may be substantially more traffic between data centres as model pull data from the Internet and elsewhere for training purposes, but this will flow across high-capacity fibre connections which can be expanded easily if needed. At present user AI interactions are generally in the form of text, which amounts to miniscule amounts of traffic. Indeed, if time is diverted from consuming video to AI interaction, then AI may reduce the amount of network traffic. In future, AI interaction may involve more in the way of images and video, but these are generally slow to be generated and so total traffic is again very small. The occasional AI-generated video of a few minutes length will make no difference compared to the amount of video already consumed. Even if AI were to generate substantial video content it is likely that consuming this would just replace other content and not add to overall access network traffic.

AI might enhance networks. It already helps network operators with customer care, fraud detection, security and preventative maintenance. Some believe it could be used to learn how best to optimise each individual cell in a mobile network for greatest capacity. However, there are many obstacles to be overcome before this can occur including finding ways to train the system and ensuring that it does not result in network failure through unforeseen errors or unanticipated interactions with other intelligent optimisers within the network. And if mobile data growth is slowing then there is limited benefit in enhancing network efficiency.

## 4.5  IoT and digital twins

IoT is growing steadily and there are many use cases that could bring significant benefits across a range of industry verticals – which has been the case since about 2010 when the initial enthusiasm for IoT first developed. As I set out in "The IoT Myth", the key constraint on IoT growth is not connectivity but the willingness and ability of organisations and industries to embrace the change in working practices needed to realise the benefits from IoT solutions. This ability is unlikely to change, and hence 5G or 6G will not materially affect IoT growth. Indeed, as mentioned, despite aspirations for IoT to be one of the three pillars of 5G, in the end 5G simply subsumed NB-IoT from 4G as its main IoT connectivity. It has since developed "RedCap" – a solution that has less functionality than 5G but more than NB-IoT, as a way to save costs over a full 5G IoT modem. The demand for such a niche product is very unclear.

The latest hope is that digital twins will somehow invigorate the IoT marketplace. A digital twin is broadly a simulation of a working system such as an airport, where the simulation is fed real-life information from sensors and hence can predict forwards to future states based on a good understanding of the current state. Digital twins can be used in multiple environments, but the data flows from the sensors needed are generally minimal – sensing number of people, temperature, number of items (eg trolleys) and so on. Such sensors send very little data – just a few bytes for each reading. Even with a very large number of sensors total traffic volumes are tiny.

The predicted 6G use cases mentioned above tend to stress that digital twins will need low latency. This seems unlikely for most applications – for example a twin of an airport is unlikely to need information more quickly than within a second or so. For some fast moving or rotating machinery, such as the jet engine mentioned, then some sensors may need to activate controls extremely rapidly, but there are few jet engines in the world, and it is likely that a proprietary short-range technology would be used within the engine or from the engine to the airplane.

For IoT to generate material levels of traffic it would need to be sending video. But most IoT devices do not do this – they have little need for video feeds and the battery drain associated with sending video is considerable. There are possible use cases for security cameras – and indeed many home security cameras link via Wi-Fi to send video. They only send video when an event triggers them such as detecting a person nearby.

The charts from Transforma[37], below, are useful in gaining a sense of perspective.

---

[37] https://transformainsights.com/research/forecast/highlights

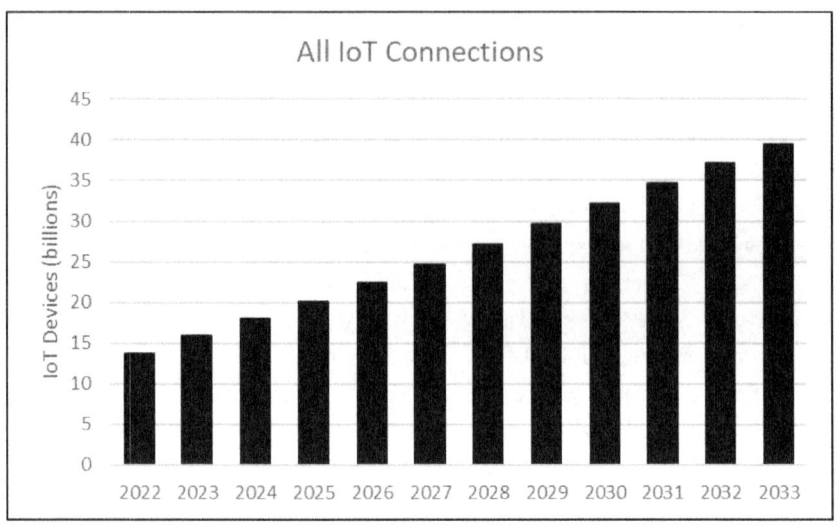

Figure 4-2 : IoT connections forecast: Source Transforma Insights

Overall IoT connections are around 18 billion – about 2 for every cellphone. Connections are forecast to double over the next ten years. But the chart below shows that few of these are on cellular networks.

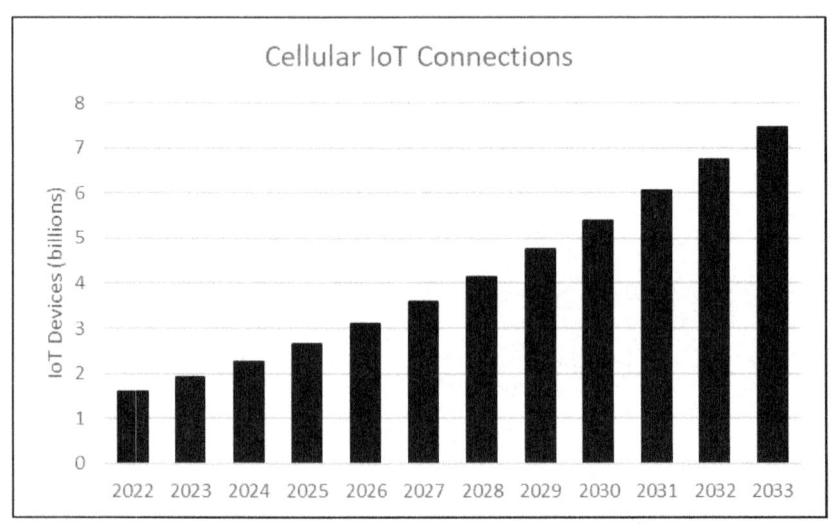

Figure 4-3 : IoT cellular connections forecast: Source Transforma Insights

Is there anything that could reignite growth?

Only about 2 billon devices are currently cellular-connected with numbers forecast to rise to about 7 billion by 2035. To put that into perspective, by 2035 there will be around one cellular-connected IoT device per cellphone. The vast majority – over 80% - connect via other technologies such as Wi-Fi, Bluetooth and private IoT networks running LoRa and similar. To deliver growth that would make a difference each cellular-connected device would need to be transmitting over an hour of video per day. Yet we know that the vast majority send only low levels of data.

IoT is sufficiently mature and well-understood that significant changes in growth and usage levels are unlikely. IoT has not shown a need for 5G levels of performance, with NB-IoT being more than sufficient for most applications. That 5G or 6G would bring about a major change in this trajectory seems very unlikely and talk of digital twins is just a "fashionable" way of discussing monitoring applications that have long existed.

## 4.6 FWA

Fixed wireless access – using mobile networks to deliver home broadband – is an area that could lead to huge data growth on mobile networks. Historically, FWA has been used to provide limited broadband to those who have no, or very poor, fixed connections. However, cellular coverage in areas of poor broadband is often weak, so the scope for mobile to fill this gap is limited and should shrink as fixed broadband is progressively rolled out to more rural areas.

More recently MNOs have started to market FWA to those who have good quality fixed broadband. The FWA offering may be cheaper or may appeal to those who are renting and want a more flexible subscription that they can "take with them" as they move. It can also be beneficial in apartment blocks where the fixed broadband comes into the building at a single point and then is distributed around the building, often suffering reduced data rates or reliability as a result. This offering appears to be successful in a few countries including the US,

Austria and New Zealand. In the US it now holds around 7% of the broadband market[38] with predictions that this might double by 2028[39].

This is not data growth as such, it is a transfer of data from one network to another (unless there was poor fixed broadband before and FWA improves this, enabling more traffic to flow). Overall consumption of data is not increasing as a result. However, the effect is to materially grow traffic on mobile networks (and shrink it on fixed networks). With each home generating 10x to 50x as much as each mobile subscriber, a small percentage of new FWA customers can result in a large overall data growth.

Conventional wisdom suggests that FWA should not succeed. A good quality fixed connection is faster, more reliable and has a lower cost of provision than a mobile connection. Fixed should win out over mobile.

What appears to be happening is that MNOs are using spare capacity on their mobile networks to deliver FWA. As such, they can offer FWA almost for free as the capacity is available and would otherwise go unused. And they may even add capacity where they can add extra carriers at little incremental cost – within their existing spectrum allocations and in the same bands as already deployed. And FWA only seems to be successful in countries where there is a monopoly incumbent fibre provider who charges relatively high fees for access.

If this hypothesis is correct then FWA will increase mobile data growth, but it will not require any additional infrastructure build since the MNOs will cap their FWA offerings at the level of their network capacity. The overall conclusions drawn in this book will remain correct and mobile data growth will stall around the same time as expected albeit with a "blip" on growth around now as FWA fills up networks. Or alternatively, the fixed network provider will eventually react, lowering their prices to stop losing customers. Since their costs are lower than for mobile network operators then once they react, they ought to be able to win back customers.

---

[38] https://www.telecomtv.com/content/access-evolution/fwa-takes-near-7-share-of-us-broadband-market-49879/
[39] https://www.globaldata.com/media/technology/fwa-technologys-share-grow-nearly-16-us-broadband-services-market-2028-forecasts-globaldata/

Is there anything that could reignite growth?

If this hypothesis is incorrect and MNOs find it profitable to add new frequency bands (which they will need to buy at auction) and potentially new technologies to their networks and can continue to grow the FWA subscriber base perhaps to 50% or more of broadband connections, then this would change overall conclusions. Data growth on mobile could return to 2015 levels and 6G or similar would be needed.

I continue to believe that fixed broadband can generally be more cost-effectively delivered via fibre, FTTC, cable or similar and that FWA can only compete where it is acting to fill capacity and/or where the fixed providers are charging excessive "monopolistic" fees. But I confess to being very surprised as to the penetration levels already achieved in the US.

## 4.7 Autonomous cars

The idea that a key application for 5G was autonomous cars has somewhat faded now, and this section need not be overly long. 5G proponents postulated that autonomous cars would be under centralised control and would need highly reliable and very low latency communications in order to perform well. This could generate vast amounts of data and a strong need for ubiquitous 5G deployment.

As I and others pointed out at the time, this clearly could not be so. No autonomous car could rely on 5G since there would be times when 5G was not available – in rural areas and tunnels for example. If the autonomous car could not function in these areas it would have limited value. Hence, autonomous cars would need to be "autonomous" – they would need to work well without network connectivity. If they could do that, then there was little need for connectivity, other than for updates on traffic conditions and similar. This has been repeatedly confirmed by Tesla and others, publicly saying that their cars needed no connectivity, although where there was connectivity it would be used where appropriate.

We do now have a limited number of autonomous vehicles – mostly "robo-taxis" in US cities. They do not need low latency 5G.

Perhaps more importantly autonomy has proven a far harder problem than many foresaw. Back in 2016, Elon Musk and others were predicting imminent full autonomy. The consensus view in 2024 is that it will take at least another decade for full autonomy to be widespread across many geographies, and some experts doubt it will ever happen except for specific routes and geo-fenced areas. It is clear that there will not be a large number of autonomous cars generating large amounts of telecoms traffic any time soon, if ever.

More likely is partial autonomy in those areas where it is easier to implement. Motorways, dual-carriageways and roads outside of towns and cities are much simpler for autonomous vehicles than built up areas. It is quite plausible that in the next few years cars will be able to take over once they have been driven to one of these roads, handing back control to the driver as they approach more complex areas. During this time the driver will likely want to be entertained or be productive, both of which might involve video. Hence, data consumption in the more rural areas might rise somewhat as a result. Overall volumes will be small – most people will not drive a car along these roads most days so the per person increase in video traffic may only be 1-2%. In the rural cells more capacity may need to be added, but typically these cells have significant spare spectrum[40].

## 4.8 Paying for growth

Some have suggested that growth might occur, for example, from cars just continuously streaming video feeds which might be useful, perhaps for autonomous car training purposes. Or that we might all wear bodycams which just streamed video, because they could. But someone has to pay for any new data usage – so there needs to be sufficient benefit from the new service. If there were an application that generated a lot more traffic, then this would lead to cost for the MNOs who would need to add extra capacity into their network. If the data was coming from new devices, such as cars, the MNOs would want to

---

[40] Although it is possible that (1) this spare spectrum may get used for FWA and (2) the cells run much less efficiently because many more users are at cell edge and so consuming more of the spectrum resource.

charge for the connection and data usage, likely at a rate commensurate with that paid by consumers. Frivolous applications would not want to pay such rates.

Where the application runs on existing cellular phones and just adds to their consumption then this becomes harder for the MNOs to charge for given that they already offer large data buckets that are likely sufficient to accommodate a lot of growth. But "fair usage policies" might be invoked to curtail applications that consumers do not value much but generate a lot of data. And hungry applications typically involve a lot of battery drain, which consumers will not tolerate.

The point here is that it is not sufficient to identify applications that might generate data, they need to generate valuable data for which there is a willingness to pay. For applications such as IoT there is often a very limited budget for the cost of connectivity meaning that designers will be cautious as to the amount of data that their solutions are consuming.

## 4.9 A solution in search of a problem

While the last chapter suggested that there was very strong reason to believe that data usage was going to plateau and that mobile and fixed networks already delivered all of the speed and capacity that we needed, this chapter has sought to test that forecast by looking for applications that might increase data usage or require speeds or other features from networks that are not currently being provided.

Proponents of 5G and now of 6G continue to seek problems that will justify the solution that they are developing. This has not worked well to date with 5G where no new applications have emerged at scale, and the predictions made in 2016 look increasingly fanciful. However, stakeholders such as manufacturers and politicians continue to "double down", proposing ever-more outrageous applications and suggesting that the reason these have not emerged to date is because 5G networks are insufficient.

This chapter shows that the applications suggested are highly unlikely to change network requirements. The reasons are varied – some are unlikely to emerge, others will run on Wi-Fi, some will have very low levels of data usage because of small numbers of devices (relative to the number of cellphones).

The biggest unknown is FWA. If FWA subscriptions continue to grow, then this will add very material loading to networks. My belief is that FWA will grow to take up available network capacity, and capacity that can be added relatively cheaply such as deploying additional carriers in spectrum already owned. However, it seems likely that it will not be economic to invest in more capacity beyond this by either acquiring more spectrum or building more cells to handle further subscribers. As a result, FWA will not increase spectrum requirements on networks and not materially change investment levels.

I now turn to look at the implications across the industry of a world where networks are sufficient.

# 5 Implications

## 5.1 Introduction

Ending 100 years of development and growth in telecommunications technologies and networks has huge implications. For example, much of the mobile telecoms industry has adapted to, and become dependent on, the 10-yearly cellular generation cycle. As discussed below, this is no longer needed. The pain that this change will cause the industry is already visible in job cuts across the entire ecosystem from manufactures to operators to consultants.

Although the transition from growth to utility will hurt, the future remains bright. Connectivity is essential for everyone and will remain so, likely forever, especially with connectivity increasingly providing both critical economic and social functions. This means provider revenues are assured and that key players will be supported through difficult times. Utilities are generally good businesses as long as they sit within a supportive regulatory framework and remain able to deliver sufficient return to their investors. As seen in many places already, governments see the merit of investing in connectivity infrastructure to deliver economic and social benefits and to avoid social divides.

This chapter looks at the implications of the end of telecoms history across the ecosystem. It starts by considering future generations of technology since the decisions made here will have major implications for suppliers, operators and others.

## 5.2 6G and Wi-Fi8

Mobile connectivity advances through releases of new technology. For cellular this is the generational updates of which 5G is the latest. Wi-Fi has a similar approach with Wi-Fi7 being the latest generation. The generational approach helps ensure harmonisation and interoperability as well as coordinating research, spectrum release and political interest.

# The end of telecoms history

As discussed earlier, the key focus of new generations of both cellular and Wi-Fi has been better performance (speed and latency) and increased capacity (more spectrum and greater spectrum efficiency).

As we move to a world of sufficiency, there is little need for either better performance or more capacity. This would suggest that there is no longer a need for new generations.

The generational approach could be re-purposed towards what is needed. As discussed in the next chapter, key among this is better coverage. A 6G standard that enabled rural and in-building coverage to be delivered much more cheaply would be useful. Other changes that MNOs and others are interested in include reducing the energy consumption of networks and increasing the ability to automate their operation. Better sharing across networks and network-of-network or "hetnet" operation would also be valuable.

What is not needed is increased capital and operational cost for operators and others without delivering increased revenue streams - as was the case for 5G. We do not need a repeat of 5G.

It seems quite likely that the generational super-tanker and its associated standards activities cannot be turned from its current path, and that both 6G and Wi-Fi8 will contain much that is not needed. If so, 6G will be costly for MNOs and they may decline to deploy it. For Wi-Fi8 there may be little cost incurred other than slightly more expensive routers which consumers and businesses will buy when they decide they need to update their Wi-Fi networks.

We will likely still need standards. Even if there is no need for faster networks, there will need to be changes, for example in response to security concerns or to a changing market structure with more network sharing. The work of 3GPP and the IEEE will continue, but at a much lower level of activity.

There is less of a generational approach with fixed networks. New standards for passive optical networks (PONs) occasionally appear but there are already standards that deliver much faster rates than routinely deployed. Upgrades seem unlikely given that even the lowest rates available are 1Gbits/s or higher.

Implications

## 5.3 Manufacturers

The implications of the end of telecoms history are most obvious in the equipment supply industry – companies such as Ericsson, Nokia, Samsung and others. If networks are not expanding to deliver more capacity and new generations of technology are not needed, then network operators will only buy equipment to replace obsolete systems. While this is still a large market, it is smaller than the one that the major suppliers are accustomed to and will require some adjustment. This adjustment has already started with, for example, Ericsson and Nokia both announcing redundancies in the last few years and smaller suppliers, such as Airspan, going through bankruptcy. Quite how much the market will contract is hard to predict. Operators have already been cutting back with less investment in 5G equipment since 2022 than in the previous few years and are likely already close to "maintenance only" spend. But if there is increased network sharing, especially of active equipment, then instead of one base station per operator per mast we will only have one base station for all operators per mast further shrinking the equipment supply market.

We might, then, expect perhaps a few more years of contractions in the RAN market before it stabilises. Hence, depending on factors such as sharing we may see, at the most optimistic, only small falls in the RAN market (albeit coming on top of a few years of falls already) or at the most pessimistic something like a halving of market size as sharing reduces the need for equipment.

With a smaller market, suppliers need a lower cost base. How they get there is up to their management to decide but there are some areas where they clearly could make savings. Equipment suppliers will not need to spend as much on R&D which will be focused more on improving their current products than inventing new generations of technology. Nor will they need to attend so many standards meetings or worry so much about filing IPR. Reductions in these areas can help better match their costs to future market revenues. The cost per base station (and other products) will also likely have to rise to reflect the fact that the indirect costs are now being spread across a small volume of sales.

Overall, the future is assured for the larger suppliers such as Ericsson, Nokia, Samsung and Huawei albeit at a lower level of sales than currently. The smaller suppliers such as the O-RAN companies like Mavenir and Parallel Wireless might struggle more, as they try to grow to a profitable level in a falling marketplace.

I have not discussed O-RAN much. Broadly it is the concept of opening more interfaces in the RAN in order to allow operators to buy different parts of the RAN from different suppliers. In particular, the base station can be separated into hardware and software, with different suppliers for each. Openness is generally a good thing, and the O-RAN concept is a sound one. If O-RAN had arrived before 5G it might have found a large market. However, with RAN sales falling and MNOs heading towards replacement purchase it becomes very hard for O-RAN to gain significant market share. Generally, an MNO replacing some obsolete equipment will find it easier to replace it with equivalent equipment, probably from the same supplier. The benefits of changing to O-RAN in such a world are small. And more generally, the need for supplier diversity is reduced where the supply is less critical – occasional replacement rather than the implementation of more advanced solutions. All suppliers will struggle, O-RAN suppliers especially so.

## 5.4 Operators

Most mobile and fixed operators have not seen revenue growth above inflation for many years but hold out hope that somehow this will turn around. It won't. Operators need to accept that they are important utilities, delivering data connectivity reliably. They should ideally restructure and cut costs to adjust to this new reality.

Indeed, there is scope for new low-cost operators to emerge that outsource much of their operations and adopt a similarly disruptive model to the low-cost airlines. For example, imagine at the most extreme a mobile operator that:

- Had no masts, renting space from a TowerCo.
- Had no RAN equipment, renting it from the suppliers along with a maintenance contract from them.
- Had no core network, buying it as a service from a hyper-scaler.
- Had no shops or physical presence, performing all activities on-line.
- Had no central office, using rented premises.

- Potentially, had no direct customers, selling wholesale capacity to MVNOs who handled the customer relationship.

Such an operator would have as an asset their spectrum licence and other rights to be an operator and their brand. They would essentially be a project management entity, with perhaps only a few hundred staff. This would give them a dramatically lower cost base than their competitors and the ability to offer lower cost tariffs as a result. Since consumers mostly select their operator on the basis of cost, they might win a larger share of the customer base and become increasingly profitable (at the expense of the other MNOs). This strategy is explored further below.

More fundamentally, many MNOs are currently making insufficient returns. Once history has ended they would like to be in a place where they are suitably profitable, with returns at least above their costs of capital. There is no reason why they should not achieve this since it is likely that consumers would pay much more for mobile connectivity if they had to – and indeed consumers in the US, for example, currently pay much more than those in Europe. Reducing competitive intensity would allow better returns at a cost of higher consumer prices. This is a controversial point given that regulatory and competition authorities tend to focus on low rather than competitive retail pricing as an indicator of a functioning competitive market and are often opposed to consolidation that might deliver better operating economics for the industry for fear of raising consumer prices.

In fact, it would be possible to have more profitable MNOs and lower consumer prices if the MNOs' costs fell. Since MNOs do not need to invest so much in infrastructure to expand capacity or deploy the next generation then their costs ought to be lower. And with a move to utility status it will be much easier to correctly forecast network investment required and future revenue streams[41]. As discussed in the next chapter, greater network sharing might also be an appropriate response to sufficiency but lack of ubiquity. This will further reduce costs and also competitive pressure with operators becoming closer to MVNOs.

---

[41] As I have argued, MNOs have been overly optimistic as to future revenue streams associated with new technology in recent years.

The path to a future where MNOs are suitably profitable (able to maintain networks, but not making excess returns) will vary from country to country and the end destination is unclear.

**A logical MNO strategy**

While the whole industry – all the MNOs – could adapt as set out above, becoming more profitable through reducing investment and increased network sharing, it is possible for one MNO to "break ranks" and try to gain competitive advantage. For example, a far-sighted CEO could:

- Repurpose the MNO towards low cost and enhanced coverage rather than "best 5G" or "fastest network".
- Reduce network costs by minimising investment and optimising a stable network for lowest operational costs.
- Improve coverage and off-load traffic by building a Wi-Fi aggregator and using HAPs to deliver better, cheaper rural coverage.
- Build sustainable competitive advantage to gain subscribers and hence grow profitability.

The aim would be to have a lower cost base than other MNOs, be able to reduce subscriber prices as a result, gain additional subscribers (and hence further increase profitability) and manage the traffic that results through off-load. This is now explained in a little more detail.

To reduce costs, MNOs can:

- Limit infrastructure investment by not upgrading networks to 5G unnecessarily and only adding more capacity where there is severe congestion, throttling subscribers a little if needed.
- Optimise the network for low maintenance – moving to single box solutions in the RAN, retiring older generations, automating maintenance and looking hard at all maintenance costs.
- Going fully on-line, removing shops and points of presence, and reviewing sales and marketing.

## Implications

- Reducing staff eg in General and Admin, R&D, deployment and outsourcing as far as possible.
- Move to a lower cost headquarters and rationalise the building portfolio.
- Consider more network sharing, including active sharing.

To improve in-building coverage and off-load traffic an MNO could set up a Wi-Fi aggregator. This concept is discussed in more detail in the next chapter, but in outline the aim is to create an entity that allows auto sign-in to existing Wi-Fi in public buildings by persuading public building owners and tenants (eg coffee shops, restaurants) to sign up. There could be a reward for users who make heavy use of Wi-Fi. The MNO can then promote the in-building coverage delivered as a result and can aggressively off-load traffic to keep data usage low on the cellular network.

To improve rural coverage the MNO should turn to HAPs and satellite – discussed in more detail in the next chapter. Tethered aerostats and similar can deliver coverage and capacity equivalent to 100 cells for about the cost of five. An MNO could partner with a company with patented HAPs technology for exclusive in-country rights and use this to both expand coverage and overlap expensive cell sites which can then be decommissioned. If enough expensive existing rural cells can be turned off as a result this may be a cost-neutral venture. The near-ubiquitous coverage provided then offers strong options for differentiated marketing.

MNOs can then experiment with low-cost brands set up alongside their existing brand – as Telefonica O2 do with GiffGaff in the UK. These brands might have no shops or customer support, relying on FAQs. AI bots and customer forums to resolve problems. They might be eSIM only (so no subsidised handsets) and have realistic upper traffic limits such as 10 and 20GBytes/month, all for a much lower cost. If these brands became successful, the "full service" model could be retired.

This approach works because lower costs allows lower tariffs. That translates to more subscribers which means more revenue. The increased traffic can be managed by:

- Removing heavy users.
- Wi-Fi off-load.
- Working with key traffic generators to compress video etc.

Sustainable advantage can be achieved if the MNO can:

- Own the Wi-Fi aggregator or have preferential access.
- Develop a clear "lowest cost but best coverage" positioning, delivering brand differentiation.
- Gain exclusive access to HAPs by funding final development of the optimal solutions, and hence having lower costs for rural coverage than competitors.
- Being the best at cost-cutting, especially automation of network operations.

This is a quite different approach from many networks which advertise "most extensive 5G" or "fastest network". But that strategy has not worked – for all the reasons set out in earlier chapters. The MNO that delivers what consumers actually want can gain many years of competitive advantage over those MNOs that continue to fight the battles of previous decades.

## 5.5 Academics

Academics have provided the underlying research that has led to many advances in telecommunications. Their role in basic research, often without a clear end benefit, is essential.

In the last 10-20 years they have become coupled more closely with cellular generations. Funding from sources such as the EU Horizon programme has often been directly coupled to 5G and now 6G. Manufacturers have provided funding towards university research centres established often with a clear and specific focus on the next generation of mobile technology. If there is less need for future generations of mobile then these sources of funding will reduce, potentially to nothing.

This may not be a bad thing in the long run. Having research funded by entities with a specific interest in furthering the next generation of mobile leads to conflicts of interest. It has also resulted in academics moving into areas that perhaps they are less qualified to address, such as the likely future demand for particular technologies, and was certainly one of the elements that led to the over-hyping of 5G.

A better model for funding is needed. This model would allow academics to research whatever areas they felt were of interest. Operators and others should set out clearly what their priorities are which would likely influence what academics chose to research. Future funding models would likely draw upon the generic research funding frameworks that exist to fund academia. But these existing models need to understand that there is a funding gap caused by industry reducing their research activities and hence that there may need to be an increased budget from government as a result.

Academics should continue their research – they may find cheaper or better ways to deliver existing services. But their focus should be systems that deliver real user benefits in a world where we have sufficient communications. These might be lower cost and more environmentally friendly solutions, or better ways to deliver rural coverage, rather than ever-faster systems.

## 5.6 Regulators

In the longer term – once data usage has plateaued - regulators will no longer need to find new spectrum bands for cellular every few years and then conduct auctions. Indeed, the demand for spectrum may abate across most users. A different way to regulate spectrum will be appropriate, discussed further in the next chapter, to deliver quality national networks rather than to use markets to optimise the use of spectrum. This may require more effort than regulators currently expend on telecommunications.

In the shorter term, there may still be a need to find new spectrum. I have predicted that growth rates will fall from around 20% per year now to zero by around 2027. Cumulatively that is still a growth of 20% in network traffic. Ericsson's numbers are much higher, with their assumption that growth does not plateau until the 2030s and their numbers suggest a cumulative growth of around 100%. Even on my numbers, if MNOs were fully using all their existing spectrum in congested areas and wanted to avoid building more cells then they would need around 20% more spectrum than they currently hold. In the UK, for example, collectively the MNOs hold some 1400MHz of spectrum. A further 20% would require another 280MHz of spectrum to be found. If my numbers are low, then much more could still be needed.

The picture is unclear. Many MNOs have likely not fully deployed all the spectrum they gained in 5G auctions and may be able to make further gains through re-farming any spectrum still used for 2G and 3G. If they are close to meeting demand and do not expect much further growth then they may decide to throttle demand slightly, or look to ways to reduce demand such as compressing video. Some MNOs are suggesting that they will need the upper 6GHz band, discussed below, but beyond that they do not expect to need more spectrum. Hence, there may be one last spectrum award to undertake before 2030.

Regulators may also have to consider whether fewer operators may be better for a country, with perhaps only a single underlying fixed and mobile network in many places – just as we only have single network for electricity, water, gas, sewerage, rail, road and other utilities.

More broadly, regulators should set out their thinking on the ideal structure of the telecoms networks and markets at the end of telecoms history where priorities are no longer faster and higher capacity, but cost, reliability and ubiquity.

## The upper 6GHz band

An example of how thinking needs to change is the upper 6GHz band of spectrum. Both the cellular community and the Wi-Fi community claim that they need access to this band to handle "ever-growing traffic levels". As we have seen, it is not clear whether this is so. There is still some growth in traffic – between 20% and 100% in total before growth reaches a plateau in around 2027 – so it is possible that there will be some need for more spectrum, but equally many MNOs are not using all of their existing 5G spectrum and most Wi-Fi deployments are not fully using the lower 6GHz band. It is likely that neither community has a compelling need for the upper 6GHz band. Equally, there is no benefit in leaving the spectrum vacant[42].

The classic regulatory approach to assessing the best use for a band of spectrum is to look at the case for both competing uses, ideally on an economic basis, and then make an "evidenced" decision. That will not work here where the evidence suggests limited economic benefit from either use. A better approach, amidst this uncertainty, is to allow the different users to share the spectrum and to observe what uses emerge over time. If one user does turn out to have a strong need then sharing rules can be modified towards their use. If neither does then, in due course, other users can be allowed into the band.

This approach has been suggested by Ofcom[43] with some sensible ideas as to how sharing could occur. The role of spectrum regulators in the future may be more about sharing than about auctioning.

## *5.7 Politicians*

Political intervention in communications has grown over time. In part, this is due to geopolitics as explained in Chapter 2. Politicians typically demand that their country be high in the league tables of whatever is the latest technology – homes

---

[42] It does have some incumbent fixed links and satellite use, so it is not completely vacant, but there is room for much more use in most countries.
[43] https://www.ofcom.org.uk/spectrum/innovative-use-of-spectrum/hybrid-sharing-to-access-the-upper-6-ghz-band

passed by fibre or population covered by 5G. They also tend to want industrial strength in the latest technology and ideally self-sufficiency.

These desires are understandable, and communications is a critical national resource with large societal benefits. Politicians should have an interest and should set out national objectives. But going forward these objectives need to change. More critical than having the latest technology is the percentage of homes that have sufficient broadband and the percentage of the country that has adequate mobile coverage. Politicians need to be less focused on simple headline numbers such as "gigabit connectivity" and instead understand somewhat more subtle measures.

The next chapter discusses in more detail the role politicians should play in directing subsidies towards delivering a ubiquity of sufficient broadband and mobile coverage.

One area politicians should broadly not get involved in is supply chains. There has been much rhetoric recently about reducing reliance on suppliers from countries that are not considered friendly and building greater domestic supply capabilities. This has led some, such as the US and the UK, to promote open RAN (O-RAN) in the belief that the open interfaces this promotes will make it easier for new entrants in the equipment supply market to emerge.

Interventions in O-RAN have not been particularly successful with companies struggling in the declining market for infrastructure. Increasingly MNOs will only be buying equipment to replace aging base stations and other network components. This is not a critical supply area. If items are temporarily unavailable this will not be an issue, nor is there any need to have the latest technology since users will not benefit from faster networks. Money and effort targeted at ubiquity is far more appropriate than at developing national capabilities for a market that only needs to maintain existing networks.

Implications

**The EU white paper**

As a case study it is instructive to look at the recent EU white paper "How to master Europe's digital infrastructure needs"[44]. The paper starts out by making the case for advanced communications networks saying:

> Without advanced digital network infrastructures, applications will not make our lives easier, and consumers will be deprived of the benefits of advanced technologies. Only with the highest performance of such infrastructures, for example, will doctors be able to care for patients at a distance rapidly and safely, drones be able to improve harvests and reduce water and pesticide use, while connected temperature and humidity sensors enable real-time monitoring of the conditions in which fresh food is stored and transported to the consumer

As I have shown earlier this is mis-guided. Looking at their example applications:

- Remote surgery will be performed using fibre optic connections not over 5G, and in any case is a very low-volume application (and increasingly the laughingstock of the ludicrous 5G applications proposed).
- Drones fly today using Wi-Fi and 4G. The key issue for many is coverage, especially in rural areas, not the need for higher speed or lower latency, all of which are more than adequately delivered with 4G and Wi-Fi.
- Temperature and humidity sensors require minuscule bandwidths and typically can handle latencies of minutes. NB-IoT and LoRa are good solutions to their needs and available today.

The EU then goes on to claim that:

---

[44] https://digital-strategy.ec.europa.eu/en/consultations/consultation-white-paper-how-master-europes-digital-infrastructure-needs

> Advanced digital network infrastructures and services will become a key enabler for transformative digital technologies and services such as Artificial Intelligence (AI), Virtual Worlds and the Web 4.0, and for addressing societal challenges such as those in the fields of energy, transport or healthcare and for supporting innovation in creative industries.

Again, this is misguided:

- AI, as shown earlier, works on the slowest of communications networks – key is the models and the processing power needed to train them.
- Virtual worlds (ie the metaverse) have been shown to be a flawed vision and very unlikely to be material, especially given VR headset sales, but in any case can be delivered well on Wi-Fi and fixed broadband.
- Web 4.0 is very ill-defined, but can likely be delivered with existing networks, perhaps with some private Wi-Fi and 5G deployment.
- Societal challenges are very real but generally best addressed with IoT solutions that can monitor and then cause appropriate action such as observing a patient in their home and triggering an alarm if their behaviour changes. This is achieved with NB-IoT and Wi-Fi.

The EU white paper is a very weak effort at making the case for "advanced digital networks" – understandably so since there is no case to be made.

The paper then lamented the failure of the industry to hit the EU's targets. On the fixed side it notes:

> As regards fibre coverage, progress beyond 80% by 2028 does not seem likely, putting the achievement of the 2030 target of 100% in doubt. In comparison to the 56% fibre coverage in the EU in 2022, the US, which has traditionally relied on cable, had 48.8%, while Japan and South Korea each reached 99.7%16, due to clear strategies in favour of fibre.

As I have shown, there is no need for FTTH. As long as data rates above around 50Mbits/s are provided then no further benefits can be delivered with faster rates. Reaching 80-90% fibre penetration is probably sensible – the remaining 10-20%

is likely better served with FTTC, FWA and other solutions. A target of 100% is not sensible and likely to be hugely expensive. International comparisons are irrelevant – or at least the comparison should be percentage of homes with at least 50Mbits/s and not percentage with fibre. Just because another country has pursued an inappropriate strategy does not mean the EU should attempt to better it.

On the mobile side the EU says:

> As regards 5G roll-out, while basic 5G population coverage in the EU currently stands at 81% (with only 51% coverage of the population in rural areas), this metric does not reflect the delivery of actual advanced 5G performance. Most often, where 5G is deployed, it is not "stand-alone", i.e. with a core network separate from previous generations. Prospects for deployment of 5G stand-alone networks ensuring high reliability and low latency, which are key enablers for industrial use cases, are not good. The deployment of such networks can be estimated at significantly less than 20% of populated areas in the EU.

As I have shown in "The 5G Myth", and earlier, the only reason for deploying 5G is the addition of network capacity which is only needed in urban areas. With 81% population coverage this has likely been achieved and the current deployment is appropriate. Asking for further deployment requires unjustified spending which, unsurprising, MNOs are not prepared to make, hence the reason the EU's desire for 100% coverage is not being met.

The EU seems surprised when it says:

> On the demand side, the take-up of at least 1 Gbps broadband is very low (at 14% in 2022 at EU level) and just above half of all EU households (55%) have adopted at least 100 Mbps broadband. The take-up of high-speed fixed broadband subscriptions is lower in the EU than the US, South Korea or Japan. Standard mobile broadband take-up is better and lies at 87%, despite almost ubiquitous coverage with at least 4G networks.

As shown earlier, the reason take-up is low is because there is no need for the speed provided. The EU should assess why there is low take-up and adjust strategies accordingly, rather than lament it. And it should be pleased that there is "almost ubiquitous 4G coverage" delivering all the connectivity needed.

Repeatedly, here, we are seeing the use of a "headline" target – 1Gbits/s, fibre everywhere or 5G SA everywhere, without any assessment of whether it is an appropriate target or the costs it implies. And despite the fact that targets clearly will not be met there is no assessment why, nor willingness to change strategies in the light of strong evidence that they are misguided and clear experience that they are unachievable.

The EU then suggests that telecoms and computing networks need greater integration (rather than the "bit pipe" separation I have been arguing for here). They say:

> Convergence of European electronic communications networks and cloud services to an EU "Telco Edge Cloud" could become a major enabler for hosting and managing network virtualised functions, as well as for bringing complementary services addressing the rapidly growing markets for IoT-related products and services. This is expected to enable the transition to an industrial Internet enabling critical services in a broad range of sectors and activities of great benefit to citizens and industry alike. Concrete examples range from robot and drone services for industry, connected and autonomous vehicles interacting with edge networks deployed along the road for smart mobility and smart transportation systems, to use cases with stringent data privacy requirements such as remote patient healthcare. This calls for the broad availability of computing resources, fully integrated with network resources, to provide the data transmission and data processing capacities required by these novel applications.

It is true that applications such as remote patient care need both communications networks, such as home broadband, and computing platforms where the sensor information can be processed. But there is absolutely no need for integration between these – it brings no benefits whatsoever and increases cost and complexity. It is another example of politicians hearing a new phrase – "edge computing" in this case – and assuming it must be a good thing.

The EU then turns to look at investment in the telecoms sector, noting:

> According to a recent study conducted for the European Commission, reaching current Digital Decade targets for Gigabit connectivity and 5G may require a total investment of up to EUR 148 billion, if fixed and mobile networks are deployed independently, and stand-alone 5G-offering European citizens and businesses the full capabilities that can be offered by 5G mobile networks is deployed. A further EUR 26-79 billion of investments may be required under different scenarios to ensure full coverage of transport corridors including roads, railways, and waterways, bringing the required total investment needs for connectivity alone to over EUR 200 billion. Despite the need to densify mobile networks to achieve higher performance, EU operators are focussing on reusing existing sites for low and mid-band deployments. However, for future upgrades, e.g. 6G or WiFi 6 the required network densification is likely to increase by a factor of 2-3 by the end of the decade at least in high-density demand areas.

The EU clearly think they know better than the MNOs what is needed. According to the EU, densification (ie adding more cells) is essential. However, as shown earlier, with data usage plateauing there is no need for more capacity and no need for densification. The EU has this badly wrong while the MNOs – who can see where capacity is needed, are adopting logical and appropriate strategies. The EU does at least recognise that:

> During the last decade stocks of European electronic communications networks and services providers have underperformed in both global electronic communications indices and European stock markets. European providers of electronic communications networks and services also face low enterprise value/EBITDA multiples, suggesting lack of market confidence in the potential for sustainable long-term growth in revenues. Against this background, the proportion of at least some of the electronic communications operators' net debt over their EBITDA has

continued to grow. In addition, access to finance seems to have degraded as interest rates jumped from historical lows and widespread risk aversion linked to the new global crises result in macroeconomic uncertainty

What they fail to acknowledge is that 5G deployment has contributed to this poor financial situation. Instead, they suggest that the issue is lack of scale and that by creating pan-European operators the problems will be resolved. Discussing this strategy is somewhat outside of the scope of this book, but in passing is very unlikely to happen and even if it did to make, very unlikely to make a material difference.

I have taken a relatively long look at the EU strategy paper because it is a good example of the blinkered thinking pursued by many politicians – fixating on "soundbite" targets even when they are inappropriate and using examples of applications that are fantasies. Strategies such as that of the EU need a complete re-think at the end of telecoms history.

Indeed, it appears that the evident lack of user demand for new technology has forced politicians who cannot let go of their initial targets to reintroduce taxpayers' money into the sector to subsidise the technology and pursue goals that are now only important to them. What started as a liberalisation project to allow the market to respond to user demand has ended up with politicians saying the market must be wrong and so reintroducing public money. This is bad for taxpayers as well as bad for the operators. In fact, after some initial suspicion and ambiguity, the operators have begun to acquire a taste for subsidies. Targets have essentially meant that the whole liberalisation project is being turned on its head. Industry is now "playing" the politicians for their profit.

Politicians should instead fixate on making sure everyone has access to sufficient connectivity – 50Mbits/s broadband and ubiquitous 10Mbits/s mobile. They should stop promoting specific technologies such as fibre and 5G. The next chapter sets out how they can help the industry deliver ubiquity of sufficiency.

## 5.8 The industry as a whole

The communications industry is one that has become addicted to ever-faster, ever more complex technology. Companies, academics and others have adapted to maximise their revenue and profitability in such an environment. Few have realised that this world has come to an end. This is not because the evidence has not been there – as I pointed out it has been clear since 2016, if not earlier, that this was the case. It is more that the industry has not wanted the good times to end and so has shut their eyes to evidence to the contrary. That they have failed to turn the tide is clearly evident in their financial results and their frequent rounds of redundancies.

The future may not be bright, to paraphrase a slogan from the MNO Orange at the peak of mobile growth, but it is assured. MNOs and suppliers will be needed indefinitely, and industry revenues will stay at around current levels in nominal terms. There is plenty of opportunity for profit. It just requires those in the industry to accept the end of history and adapt accordingly. That will be painful for some but not existential.

In the next chapter I look at how we can deliver sufficient communications to all.

# 6 Delivering ubiquity

## 6.1 Coverage is not improving

While those who are well connected now have all that they need from communication networks, there are plenty who are not well connected. Issues include:

- Poor broadband in buildings in rural areas or some distance from the local exchange.
- Mobile "not-spots" or "black-spots" – small areas without coverage or with insufficient capacity for users to be able to connect as they wish– scattered across the country.
- Larger rural areas without mobile coverage.
- Trains and other forms of transport with poor connectivity.
- In-building not-spots, especially where buildings have exterior coatings that reflect radio waves.
- Large parts of developing countries.
- Those who cannot afford the cost of the devices needed or the monthly subscription.

The remaining challenge for the communications industry is how to resolve these issues such that the sufficient connectivity many enjoy can be extended to all. This chapter is about how to do this, based in part on my co-authored recent book "Emperor Ofcom's New Clothes"[45].

While coverage improved quickly in the early days of mobile networks, it has plateaued in recent years, with few operators building more coverage unless they have an obligation to do so.

At the same time, consumers are consistently saying that ubiquity is one of the most important things that they now want from their mobile device. This is

---

[45] S Temple, W Webb, "Emperor Ofcom's New Clothes", Amazon 2024.

logical – they have most other things that they want; being able to use the phone everywhere is the key remaining piece of the puzzle.

Since communications is delivered in a market economy, it may seem strange that the market is not delivering better coverage. Market theory suggests that if consumers value coverage they would pay more to the operator that provides better coverage, giving operators an incentive to expand coverage to the level that the market demands. Yet in many countries operators build approximately the same level of coverage each but stop short of what consumers would like.

One reason is that of insufficient consumer information. It is nearly impossible for most consumers to compare the coverage of the available mobile networks across the locations they frequent and places they might like to go. Coverage maps provided by operators are notoriously inaccurate and little trusted. As a result, most consumers default to selecting their operator based on the lowest priced tariff. This encourages operators to cut costs to keep prices low, and one way to do this is not to expand coverage. When consumers suffer not-spots they generally assume that the problem would be equally bad with any other operator.

There is a way to resolve this by providing an app that can track the movement of a consumer (with their permission) across a typical week or so, correlate this with the crowd-sourced coverage data of each of the available operators, and then show the consumer the percentage of time that they would be connected with each operator at sufficient quality and data rate to be able to use the apps that they routinely access. This would enable an informed buying decision that might then drive operators to compete on coverage as well as on price. I discuss how such an app might be provided later in this chapter.

Even with such an app there would likely still be insufficient coverage. For the most part this is because individuals will not value coverage as highly as society does collectively. There are externalities such as productivity benefits, social inclusion, ability to call emergency services, tourism and more that are not captured by the MNOs and hence they will tend to under-deliver. Also, there are areas such as on trains where it is too difficult, absent government intervention, to implement coverage.

Regardless of the reasons why, it is clear that at present coverage is sub-optimal, but that this situation is not changing materially in most countries. To bring about more coverage, a changed approach is needed on behalf of governments, regulators and MNOs. This section discusses what that approach should look like. It starts by considering new approaches to reducing the cost of coverage.

## 6.2 There are new tools – HAPs and D2D

Delivering coverage in the uncovered areas is generally expensive. If this were not the case then the coverage would already have been provided. These areas are mostly rural and as a result have little infrastructure. Building a mast can be difficult because:

- Planning permission can be hard to obtain in areas that are considered protected areas or national parks.
- Once obtained there may not be a road to the mast site, requiring the mast to be delivered by helicopter or similar along with all the building materials needed.
- There will not be power or backhaul, which then needs to be installed. Delivering power across many kilometres of unpopulated areas can be very expensive, as is the investment in alternative sources of energy at the site (solar or wind).
- Maintenance of the mast is hard, with visits being time consuming.

As an example, in the UK a typical mast cost in an urban area, where much of the infrastructure already exists could be in the region of $150k. But for a rural site this could rise to $1m or more. Many initiatives to deliver an increased number of rural sites have failed or been much delayed as a result of the challenges of implementation.

An alternative to the challenges and costs of delivering terrestrial rural networks might be "base stations in the sky".

There are two broad approaches to base stations in the skies – high altitude platforms which fly in the atmosphere and satellites orbiting above the atmosphere. Both are likely to have a role to play.

High altitude platforms (HAPs) represent a broad class of "flying base stations". They could be considered either as very tall telecom towers or very low altitude satellites.

There are many different types of HAPs, broadly:
- Heavier than air:
    - Conventional planes.
    - Unmanned drones, typically fixed wing eg Stratospheric Platforms, Aalto Zepher.
- Lighter than air:
    - Tethered balloons eg Softbank Altaeros.
    - Aerostats or dirigibles (balloons that have propulsion), eg Sceye.
    - Free flying balloons, eg Google Loon (discontinued).

HAPs provide a much greater coverage area than conventional cells but smaller than satellites. They tend to have greater capacity than satellites but less than a network of terrestrial cells. Coverage and capacity vary across the different types of HAPs, broadly according to how high they fly, from a few hundred meters for tethered balloons to 20km+ for some drones and aerostats.

As such, their role is in delivering coverage into areas that are too rural for cellular networks but not so deeply rural that there are almost no inhabitants. They could have a role in covering much of the world's rural areas.

The key issue for HAPs is the cost of operation. Maintaining an airborne platform is more expensive than a terrestrial mast. The most effective HAPs platform is the one that can provide much greater coverage than terrestrial cells at a cost of less than a few terrestrial cells. Different HAPs platform providers are competing on the basis that they have the lowest cost technology.

For example, for the rural areas with significant population, especially where broadband is also required, a tethered balloon, flying at a few hundred metres to 1.5 km, might be most effective. Such a balloon could stay airborne for many years, with short maintenance sessions every few months where the balloon is pulled back down, the antenna is serviced, and its helium is replenished. It might achieve a coverage of about 20-50km radius and depending on population density replacing something like 100 or more conventional cells. It could host multiple 5G base stations, using beam-forming antennas to deliver significant capacity. Of course, there are challenges, predominantly associated with planning permission but local benefits could be offered so rural communities might welcome installations, furthermore the precise siting requirements of conventional masts are removed if typical ranges are 20 to 50 km. Positioning also needs to avoid commercial airports and respect airspace change management procedures, but these issues ought to be soluble.

For the areas with less population, HAPs at higher altitude can deliver coverage across much larger areas. For example, Stratospheric Platforms' approach is a somewhat conventional plane, albeit powered by hydrogen, with a large antenna array across its underside, delivering 10s or 100s of cells across massive areas. It can be deployed almost immediately with no need for any ground-based infrastructure.

To take a particular country, delivering truly ubiquitous coverage across the UK might require something in the region of 20,000 additional conventional cell towers (per network unless shared). Such a network deployment programme could cost as much as $10-20 billion assuming two networks for the UK shared between three or four operators and would likely take decades to implement. But much of this could be replaced by a few hundred tethered balloon HAPs. Each of these might be $5 m to $7m, totalling around $1 billion. Or a mix of balloons and planes could be used, with perhaps $0.5bn spent on balloons in the areas where broadband was also needed and $0.25bn on the cost of flying aircraft to cover the other areas. Regardless, huge cost savings can be made, and solutions can be implemented much more quickly that using conventional systems.

Satellite systems providing terrestrial coverage have been present for decades, with Iridium providing coverage to bespoke handsets, and more recently Starlink providing high data rate broadband connections and Apple enabling "direct to handset" (DTH) emergency calling on iPhones with the suggestion[46] that this

---

[46] https://www.theverge.com/2024/6/10/24175479/ios-18-satellite-imessage-wwdc-2024

may soon expand to general messaging. Many companies including SpaceX, Kuiper and others are working on satellite systems that can provide direct to device (D2D) services that go beyond the iPhone emergency messaging and aim to deliver at least GSM-level service.

The key advantage of satellites is truly ubiquitous coverage outdoors (satellites tend not to work well indoors because the signal is attenuated by the building roof). Satellites could provide the final resource where terrestrial and HAPs systems are not viable. Their weakness is their low data capacity.

Hence, through a mix of conventional cells, HAPs, and satellite systems (an integrated network of networks), ubiquitous coverage might eventually prove an order of magnitude less expensive than currently envisaged using cell towers alone.

Balloons, planes and satellites are not new, raising the question as to why these solutions are only just emerging now, and why none have been deployed at any real scale. Broadly, this is because there has been little incentive for MNOs to deploy these solutions because they have not been competing on coverage, as explained earlier. Without MNO demand there has not been the market to justify the investment in developing the platforms. Only now, with increased interest in rural coverage, is it becoming viable to raise funding to develop HAPs and satellite solutions – and even now raising that funding is very challenging. However, material changes in MNO attitudes towards delivering rural coverage could lead to the "pull-through" needed to make HAPs and satellite solutions commonplace.

The next section looks at how to change MNO attitudes.

## 6.3  *Subsidy will be needed*

The reason coverage has not been improving in recent years is one of economics. The MNOs do not consider that there is a good business case to justify the costs of deploying better coverage.

As discussed, there are ways to help with this. HAPs and satellite solutions can materially reduce the cost of coverage. Apps that help consumers understand which networks have better coverage for them will provide a stronger incentive for operators to enhance coverage in areas where there are the most consumers. But even with these changes, the externalities mentioned earlier such as productivity and social inclusion will mean that insufficient coverage will be built.

The MNOs currently find spending on infrastructure challenging. Many are making poor levels of return on investment – often below the market cost of capital – and investors are unwilling to lend them more money at such poor returns. Most MNOs are reducing, rather than increasing, investment. Even if there were a good case for further investment it may be difficult for MNOs to find the funding needed from a sceptical investment community.

In fact, coverage enhancements have long been subsidised. For many years, spectrum auctions have often included coverage obligations within the licence. This is an implicit subsidy. Logically, the MNO would bid less for a licence with an obligation, reducing their bid by the anticipated cost of meeting the obligation. Governments get less auction revenue as a result. Of course, the same result could be achieved by not having an obligation on the licence, the government taking greater auction revenue and then paying the MNOs to build more coverage. The end result would be the similar, with greater transparency, but the implicit approach is easier for politicians and Treasury officials to manage.

There may not be many more auctions, as discussed in the previous chapter, but licenses that were auctioned 15-20 years ago are now coming up for renewal. Some regulators, for example as in Germany, are considering allowing MNOs to retain their spectrum in return for delivering specific coverage improvements. Again, this is a subsidy for coverage, but again it is an implicit one.

There are many other ways that implicit subsidy could be delivered. Along with my co-author I explored these in detail in "Emperor Ofcom's New Clothes". They include:

- Allowing MNOs to retain annual licence fees where they are currently paid.
- Using revenues from any auctions to directly subsidise the required network improvements.
- Facilitating ways to reduce network loading such as in-building network off-load (discussed below).
- Allowing greater network sharing, leading to savings for the MNOs which can be used for investing in coverage.
- Government subsidy for HAPs or satellite system deployment, reducing the cost to the MNO.

Alternatively, Governments could just pay MNOs to deliver specific outcomes – as they do, for example, in the UK's Shared Rural Network (SRN). In many ways this is easier to manage and more transparent but can be more challenging for governments.

The key message is that if governments desire better coverage for their citizens then they will need to subsidise it in some manner.

## 6.4 Shared networks

If delivering coverage is uneconomic, then it makes little sense to build multiple uneconomic networks. Where there is a network subsidy, implicit or explicit, then a single shared network makes more sense. This is the approach adopted in the UK with the SRN, and is effectively the case in many other countries where single shared masts are deployed and in some cases shared active infrastructure at the masts (shared radio access networks – RAN).

Sharing is the only viable option for satellite connectivity where it would not be possible for each MNO to launch their own satellite network. It also makes more sense for HAPs where the cost of a single airborne platform can be shared across multiple MNOs.

Sharing of such networks raises questions such as:

1. Who should own the networks? Should governments have a stake?

2. What are the boundaries of sharing? Should sharing be extended across the country potentially resulting in one underlying network? Should spectrum also be pooled, especially lower frequency bands?

Regarding ownership, typically shared networks are jointly owned by the MNOs that use the networks. Shared towers are sometimes jointly owned by two or more MNOs or in some cases by a TowerCo. Ownership is generally through a separate company (ie not part of an MNO) with which the MNOs have contracts for provision of shared network services. The ownership structure is probably not critical, although MNO ownership does make subsequent MNO mergers more challenging. A TowerCo model might work better.

The harder question is whether sharing should be allowed to extend from rural areas into other areas and whether we should aim for a single underlying mobile network, just as we have single networks for power, water, rail and so on.

Sharing trades off cost of infrastructure against competition. Until recently competition in mobile networks was seen as important, as evidenced by refusal by many competition authorities to allow mergers, and caps imposed by regulators on spectrum holdings. (Conversely, there has been little competition in fixed networks and the expectation is that we will end up with a single fibre connection to most homes.) However, as we come to the end of telecoms history, the argument grows that the benefits of competition - such as innovation and responsiveness to demand - will decline, and that the mobile industry will tend to a utility model. Hence, there are grounds for allowing network sharing to extend – across a whole country if appropriate – with tools such as wholesale access used to deliver some of the benefits of competition.

## 6.5 Trains

In many countries connectivity on trains is poor, with not-spots in tunnels and cuttings and networks unable to deliver the capacity needed within a carriage with many people who, often having little else to do, are inclined to heavily use mobile connectivity[47].

---

[47] Gone are the days when passengers used to read newspapers or even books!

Funding will need to be provided for a specific rail solution. The ideal rail solution is one that can deliver gigabits of data to trains using dedicated mmWave radio solutions to antennas on the carriage roof linked to Wi-Fi and cellular mini-base stations inside the carriage. This can easily be deployed on mainlines which have overhead power where the wireless solution can use the same gantries, and on other lines with dedicated small masts for the mmWave radios.

This would be a shared network – it would be cumbersome and expensive to build multiple radio solutions on the same lines. It is generally better implemented by a dedicated entity with links to the railway companies to facilitate access and safety concerns. Funding would need to be provided by government since the MNOs would likely see little revenue benefit and the railway companies generally will have limited incentives to enhance connectivity.

## *6.6 In-building*

For many, the time that they are most likely to be out of mobile coverage is when they are indoors. The building fabric attenuates cellular signals making indoor coverage challenging, especially where the building fabric includes metallised windows or similar which act as a complete block to radio signals.

The best solution to high-quality in-building coverage is to deliver the signal from within the building. This both results in better quality of signal and benefits from the building fabric reducing interference to and from other networks. In most buildings there already is a wireless network present – Wi-Fi.

Today Wi-Fi in the home and office plays a vital role in lifting a huge quantity of data off wide area mobile networks that simply do not have the bandwidth to handle it. That will continue. But visitors to people's homes and third-party commercial premises generally need a manual exchange of the password to be able to access the Wi-Fi. Visitors to many commercial buildings, especially those with well screened construction, can find themselves cut off from any mobile connectivity, but may lack password access or face other inconveniences to gain Wi-Fi access.

There have been efforts to form clubs of users willing to grant mutual access to each other's home Wi-Fi routers. But it has never been successfully scaled up into a compelling proposition. There have been good examples of cellular mobile operators creating coverage partnerships with commercial building owners. But not yet at scale. This lack of a successful and nation-wide aggregated Wi-Fi solution thwarts using Wi-Fi as a virtually free and immediately available solution to delivering near-ubiquitous in-building coverage.

What needs to be done technically looks straightforward. The barrier to using Wi-Fi is that most hotspots require manual selection of the router and entry of a password on the first time of use (after that most handsets remember these details and automatically log in). If most Wi-Fi routers were part of a scheme where the same public log-in could be used, then handsets would move onto Wi-Fi whenever they entered almost any building.

This is not a new idea, and there have been many attempts in the past, including in the UK the still-existing BT OpenZone arrangement and university EduRoam scheme. They have not gained traction partly because the technology was not there and partly because MNOs had other priorities. Technology has improved, especially with the recent OpenRoaming initiative.

In many countries, government has substantial Wi-Fi resource available in the form of networks such as EduRoam, (in Europe), GovRoam and health service Wi-Fi. Pooling this resource and enabling single one-time sign-on access will both resolve coverage issues across many public buildings and create a scheme that commercial entities, such as restaurants, offices, theatres and others will be able to join. Use of crowd-sourced data to identify the most problematic indoor coverage issues can prioritise actively targeting those organisations owning or resident in the affected buildings. A mix of publicity and customer pressure can then lead to a snowball effect where most commercial organisations will want to be part of the shared Wi-Fi network.

This relatively simple step will both deliver near-perfect indoor coverage and reduce the loading on external cellular networks.

## 6.7 A change of focus from speed to coverage

On the face of it, the solution appears relatively simple. Provide government subsidy to the MNOs to implement a shared network in areas where there is insufficient coverage, coupled with specific projects for trains and in-building.

To some degree this approach runs counter to the underlying "market forces" philosophy that has been used to manage and regulate MNOs since around the year 2000. In the market forces approach, interventions are to be avoided as they distort the market. Instead, financial incentives are used to encourage the most efficient use of resources.

Our book "Emperor Ofcom's New Clothes" explained in detail why the market forces approach was no longer working. In essence, the MNOs have little room to adapt to such forces. Regulators aim to avoid any one of them gaining an advantage, reducing the incentive to innovate. They all use the same technology, conforming to the same global standard and delivered from the same two to three suppliers. And they deliver services that are commodities where differentiation is very hard. As a result, innovation does not lead to greater efficiencies, and instead the status quo is maintained at low levels of operator profitability.

For there to be a change, governments need to set clear objectives for the MNOs (rather than assume that these are effectively set by the market). These objectives might include specific levels of coverage, perhaps measured and verified through crowd-sourced coverage information. The regulator needs to align with these objectives, perhaps adapting auctions, licence fee pricing and relevant regulatory restrictions accordingly. These are not monumental changes, and some regulators have already embarked on this approach, but for others they will require a different mindset and may need small changes to legal frameworks, especially those which the regulator works under.

There is a particularly important role for crowd-sourced data. This is the use of a subset of consumers' mobiles to gather information on network quality as the consumers go about their daily business. If enough consumers take part (and typically only around 1% of the population is needed) then the aggregated information can provide a rich insight into where there is insufficient coverage or capacity. Crowd sourcing can be used:

- In the first instance to understand where there are coverage problems and to rank these problems according to the number of people affected each day. Then, the not spots that cause the greatest problems can be addressed first.
- To verify that MNOs or shared network companies have met the obligations that they take on, for example in return for payment, by showing whether consumers now receive reliable and high-quality coverage in the areas identified as requiring solutions.
- To be part of the solution of delivering a network selection app that can tell consumers which network would provide the best service to them given their weekly movements. This app could be sponsored by government and users reminded of its existence a few weeks before their contract renewal date. Building such an app is very easy and hence the costs involved are insignificant.

## 6.8 Fixed networks

The discussion so far in this chapter has concentrated on mobile coverage. There are also challenges of broadband coverage, with buildings in rural areas often having low speeds of connectivity. Governments often actively address this with subsidies such as vouchers to cover the costs involved in trenching fibre cables across longer distances. As a result, the broadband coverage challenge is slowly being addressed.

However, it could generally be addressed faster and at lower cost if:

1. The need to deliver gigabit connectivity is relaxed. As seen, around 50Mbits/s delivers equally good service to most and gigabit rates are unnecessarily expensive – akin to requiring all cars to be capable of 200mph when the top speed limit is 70mph.

2. Other technologies are more actively pursued including FWA, HAPs and satellite, often in combination with mobile networks. For example, both fixed and mobile networks might share the same HAPs platform, delivering mobile coverage and rural broadband simultaneously.

With these changes it is then just a matter of time and money to resolve broadband connectivity issues.

## *6.9 Delivering ubiquity*

Coverage is the last challenge for mobile and fixed networks. This chapter has shown how it can be addressed, predominantly through government subsidy, but optimised using emerging technologies and realistic targets for speed. It might be considered as a postscript to the history of telecoms.

# 7 Conclusions

## 7.1 In summary

Telecoms networks have improved dramatically over the last 150 years. Initially developments focused on delivering voice calls to homes and then to mobile phones, and then from the 1990s to enabling Internet connectivity. Data capabilities have improved from 1.2kbit/s modems to networks that can deliver 100Mbits/s or more. And data usage per person has gone from a few Mbytes/s per month to hundreds of Gbytes – nearly a 1-million-fold increase. It is natural to think this progression will continue forever.

But this is not so. Beyond around 10Mbits/s on mobile phones and 50Mbits/s on fixed broadband, faster data rates make no difference to most. Other constraints such as the Internet servers that limit responsiveness. And our main use – video – only requires 3Mbits/s for high definition. Our use of data is now levelling off, with growth rates already below 20% a year and likely to fall to 0% - a flat level of usage – by around 2027.

Those who are well connected, with good home broadband and good mobile coverage – even if it is only 4G – have all the connectivity that they need. We no longer need to strive for faster networks, for more fibre or for the next generation of mobile technology. The journey that started with Morse and Marconi has come to an end. It is the end of history for telecoms.

Of course, users would like all of this for less, and they would like ubiquity, especially in mobile connectivity. This can be delivered with appropriate government targets and regulatory action.

This change from seeking ever-better networks to having all we need has huge consequences – no need for spectrum auctions, for 6G, for new standards, for most R&D, etc. Politicians should stop fixating on the latest "greatest" technology and instead be concerned about delivering ubiquity at the lowest cost. Regulators need to set aside their assumptions that they will be auctioning spectrum periodically and look instead at regulation focused on improving fixed and mobile coverage. Operators should accept that they are bit pipes and adopt a utility-like strategy and equipment suppliers need to adapt to a market where operators will replace equipment as it becomes obsolete but will not adopt new technology or expand their network. Innovation should focus on cost reduction, improved efficiency and better applications rather than capacity enhancement.

## 7.2  Satiation is a good thing

This book has set out how those who are well connected have all that they need from communications networks. Data rates on fixed broadband are generally above the 50Mbits/s or so beyond which most users will notice no difference. Likewise, mobile rates are often above the 10Mbits/s beyond which users notice no improvement. Network capacities are sufficient, and with data usage expected to plateau in the coming years further capacity expansion is not needed. We have reached the end of history for communications.

Perhaps it might have felt the same when voice communications had been delivered to all – and then along came the Internet sparking another 30 years of developments in fixed and mobile networks. Another unforeseen need could emerge, but it is hard to envisage what it might be, and very unlikely that it would happen in the next decade (otherwise we would have some early sight of it already).

Of course, users would like all of this for less, and they would like ubiquity, especially in mobile connectivity, and this can be delivered with appropriate government targets and regulatory action.

We should rejoice at this. It has been a long journey, but we now live in a world where we can call anyone from wherever we are using a small device we keep in our pocket. With the same device we can access virtually any knowledge in the world, any book, music or video content. We can order virtually anything to be delivered to our home and do so much more. Just 30 years ago this would have seemed magical and utopian. And we can do all of this for a monthly subscription that is less than 1% of average salary in most countries.

## 7.3 New thinking is needed

The outcomes and recommendations as a result of having sufficient connectivity are manyfold. This book has set out the following:

1. Politicians should stop fixating on the latest "greatest" technology and instead be concerned about delivering ubiquity at the lowest cost.
2. Regulators need to accept that there is a market failure with regard to ubiquity and look at regulation focused on improving fixed and mobile coverage and on guiding the industry to an appropriate structure for a utility business.
3. Operators should accept that they are bit pipes and adopt a utility-like strategy. Controlling costs and maintaining the existing network should be key focus areas, working with regulators on future models such as network sharing.
4. Equipment suppliers need to adapt to a market where operators will replace equipment as it becomes obsolete but will not adopt new technology or expand their network. Innovation should focus on cost reduction rather than capacity enhancement.
5. Academics should continue their research, but with less involvement in areas that they not expert in such as predicting future applications. They should take their steer from operators as to potentially fruitful areas of research.

# Index

3DTV, 24
3G, 19
3GPP, 26
4K, 24
5G, 6, 50
6G, 53, 68
6GHz band, 77
8K, 24
Aalto Zepher, 89
academics, 74
ADSL, 15
Aerostats, 89
AI, 53, 54, 55, 57, 58
Airspan, 69
Amazon, 3, 7, 9, 86
Analysys Masons, 40
Apple, 9, 56, 90
applications, 50
auctions, 45, 75, 92, 97
autonomous cars, 63
Barclays, 39
Bell Labs, 17
bit pipes, 28, 102
Bluetooth, 57, 61
buildings, 95
cable, 14
CDMA, 19
coverage obligations, 92
Covid, 45, 51
crowd-sourced, 87, 96, 97
data centres, 58
Deloitte, 8
digital twin, 54, 59

direct to handset, 90
dirigibles, 89
drones, 79
DTMF, 13
DTTV, 24
DVD, 23
edge computing, 83
*eMBB*, 7
Emperor Ofcom's New Clothes, 86, 92
energy consumption, 68
Ericsson, 9, 26, 37, 41, 43, 53, 55, 57, 69
eSIM, 73
EU, 27, 53, 57, 74, 79
Europe, 18
externalities, 87, 92
fax machine, 13
fibre, 15
Francis Fukuyama, 3
FTTC, 15
FWA, 45, 55, 61, 62, 63, 66, 99
gaming, 45, 48, 53
geopolitics, 26
GiffGaff, 73
gigabit, 32, 49, 78, 98
Google, 9, 29, 32, 89
GPRS, 18, 22
GSM, 18
HAPs. *See* High Altitude Platforms
hetnet, 68
high altitude platforms, 88
holographic, 54

HSPA, 19
Huawei, 26, 69
Internet, 13, 31
IoT, 7, 58
iPhone, 19, 35, 91
ISDN, 15
Japan, 18
Kuiper, 91
latency, 2, 7, 20, 32, 33, 34, 48, 49, 54, 59, 68, 91
linear TV, 23
location-based services, 29
Loon, 89
LoRa, 61
manufacturing, 53
Marconi, 12, 106
market forces, 97
Marty Cooper, 17
Mavenir, 69
Meta, 55
metaverse, 55, 80
MIMO, 25, 28
*MMC*, 7
MMS, 19
mmWave, 95
modem, 14
Morse, 12
Motorola, 9, 16, 17, 106
NB-IoT, 7, 58, 61
Netflix, 23, 31
Nokia, 9, 42, 55, 69
Nortel, 9
not spots, 86, 98
Ofcom, 43, 77, 86, 106
OFDM, 20
O-RAN, 69, 78

OTT, 9, 28
Parallel Wireless, 69
picture messaging, 29
politicians, 2, 16, 27, 65, 78, 92
PONs, 68
Princess of Wales, 17
productivity, 87, 92
push-to-talk, 29
QAM, 14
Qualcomm, 19
RedCap, 58
regulators, 43, 50, 75, 76, 97, 102
remote control, 7, 48, 51, 52
remote surgery, 6, 79
roaming, 18
SA, 51, 82
satellites, 88, 89
S-curve, 37, 39, 40, 41
security, 17, 18
Shannon, 25
shared network, 93
smart glasses, 57
SMS, 18
SNR, 25
social inclusion, 87, 92
Softbank Altaeros, 89
SpaceX, 91
spectrum, 18, 19, 20
SRN, 93
Stratospheric platforms, 89
Stratospheric Platforms, 90
streaming, 23
subsidy, 92
surveillance, 51, 52
tactile communications, 48
TCP/IP, 32

# Index

Tesla, 63
The 5G Myth, 7
The IoT Myth, 58
tourism, 87
trains, 53, 87, 94, 95, 97
Transforma, 59
undersea cables, 12
*URLLC*, 7
US, 18, 19
utilities, 27, 29, 70, 76
VCRs, 23
VDSL, 15
verticals, 58

video, 31, 37
video blog, 57
video calling, 29
video calls, 52
VisionPro, 56
VoLTE, 20
VR, 53, 55
Web 4.0, 80
WhatsApp, 28, 29, 52
Wi-Fi, 52, 67, 95
XR, 53, 55
ZTE, 26

## About William Webb

William is an independent consultant providing advice and support to a wide range of clients on matters related to digital technology.

He was CTO at Access Partnership, one of the founding directors of Neul, a company developing machine-to-machine technologies and networks, and CEO of the Weightless SIG - the standards body developing a new global M2M technology. Prior to this William was a Director at Ofcom where he managed a team providing technical advice and performing research across all areas of Ofcom's regulatory remit. He also led some of the major reviews conducted by Ofcom including the Spectrum Framework Review, the development of Spectrum Usage Rights and cognitive or white space policy. Previously, William worked for a range of communications consultancies in the UK in the fields of hardware design, computer simulation, propagation modelling, spectrum management and strategy development. William also spent three years providing strategic management across Motorola's entire communications portfolio, based in Chicago. He was President of the IET – Europe's largest Professional Engineering body during 14/15.

William has published 18 books including "The 5G Myth", "Our Digital Future" and "Spectrum Management", over 100 papers, and 18 patents. He is a Fellow of the Royal Academy of Engineering, the IEEE and the IET, a Visiting Professor at Southampton University, a Board Member of the Marconi Society and a non-executive director at Motability. He has been awarded three honorary doctorates and the IET's Mountbatten medal, one of its highest honours, in recognition of his contribution to technology entrepreneurship.

Printed in Great Britain
by Amazon